SOVIET ENERGY TECHNOLOGIES

SOVIET ENERGY TECHNOLOGIES

Planning, Policy, Research and Development

ROBERT W. CAMPBELL

INDIANA UNIVERSITY PRESS

Bloomington

Library of Congress Cataloging in Publication Data

Campbell, Robert Wellington.
Soviet energy technologies.

Bibliography: p.
Includes index.
1. Energy development—Russia.   2. Energy policy—
Russia.   3. Power resources—Research—Russia.
I. Title.
TJ163.25.R9C36   333.79′0947   80-7562
ISBN 0-253-15965-2   1   2   3   4   5   84   83   82   81   80

Contents

Preface   ix

Chapter 1.   The Energy Sector and Energy Policy in the Soviet
Economy   1

Chapter 2.   Soviet Energy R and D   27

Chapter 3.   Thermal Power Generation   62

Chapter 4.   Coal Mining   99

Chapter 5.   Nuclear Power   138

Chapter 6.   Technologies at Early Stages of Development   170

Chapter 7.   Technology Transfer   203

Chapter 8.   Conclusions   231

Bibliography   251

Index   261

Tables

1-1. Soviet Fuel and Energy Production, Trade, and Apparent
Consumption   10

1-2. Energy Consumption by Using Sector, 1975   13

1-3. Fuel and Power Consumption by Major Sectors, 1975   15

2-1. Estimated Expenditures on Energy R and D in the USSR, 1975   40

2-2. Soviet Energy R and D Expenditures by Major Programs   53

2-3. U.S. Energy R and D Expenditures, 1975   54

3-1. Comparative Indicators for U.S. and Soviet Electric Power
Utilities, 1975   66

3-2. Distribution of Steam Turbogenerator Units by Size of Unit,
Regional or Utility Stations, End of Year, Percent of Total
Steam Turbine Capacity   70

3-3. Steam Parameters of Thermal Power Generating Units, End of
Year, Condensing and Heat and Power Stations   71

3-4. Distribution of Installed Capacity by Capacity of Station,
End of Year, Percent of Total   72

4-1. Comparative U.S.-Soviet Indicators for Coal Mining, 1972   100

4-2. Comparative Indicators for U.S. and Soviet Open-Pit
Coal Mining, 1972   104

5-1. Growth of Nuclear Power Capacity and Output   139

5-2. Soviet Nuclear Power Reactors in Operation and Under
Construction   140

7-1. Gas Transported on Mingaz Lines   205

7-2. Capacity Indicators for Mingaz Lines   206

7-3. Comparison of U.S. and Soviet Gas Pipeline Systems   207

7-4. Gas and Oil Pipeline, 1020mm and Above   210

7-5. Estimate of Domestic Output of Pipe, 1020mm and Above   211

7-6. Compressor Equipment on Soviet Gas Pipelines   215

7-7. Electric Submersible Pumps in Soviet Oil Production   223

Figures

3-1. Heat Rate in Thermal Electric Power Generation, Utility
Sector, United States, USSR, and Western Europe   67

3-2. Measures of Daily Load Variation in Soviet Power Systems   86

5-1. Relevance Tree for Raising the Effectiveness of Power
Production   147

Preface

This book has been conceived and written with a dual rationale in mind. It is intended first as an attempt to provide a more solid understanding than previously available of Soviet technological capabilities as a basis for interpreting and forecasting Soviet choices regarding energy options. The Soviet Union today faces a tighter energy situation and a more complicated set of energy choices than in the recent past. The managers of the Soviet economy must rebase its energy supply system to new kinds of energy resources and to new regions. They need to give a strong role to conservation as an alternative to supply expansion, decide to what degree and in what form nuclear energy can help, decide how much technology transfer should be employed as an alternative to domestic innovation to solve the associated technological tasks. Many of the potential actions the planners could take pose requirements for improvement in existing technologies or the creation of completely new ones if they are to become feasible. In dealing with this, they must make assessments regarding the probability of successful realization and relative attractiveness of alternative technical solutions to fuel problems. It is the author's hope that, by examining the status of a number of important energy technologies in the USSR and by looking at the history of Soviet efforts to upgrade technologies and to innovate in the energy sector, it will be possible to develop a much more informed perspective on the degree of success to be expected in the future in coping with technical changes, on the considerations that will guide Soviet efforts, and on the directions they are likely to follow.

There is a second important motivation for studying Soviet energy technology in detail. It is my belief that a detailed examination of concrete cases and an examination that looks at the evolution of technologies over some period of time in the setting of overall management of a specific sector add considerable concreteness and reliability to our views regarding the characteristic features of Soviet R and D management and its effectiveness. There is now a large literature on R and D, innovation, and technical progress in the Soviet economy, much of which is concerned with R and D resources, with R and D planning as an aspect of national economic planning, and with the behavior of various kinds of decision makers as they operate within the overall system of the planned economy. There are also interesting studies of technical progress and R and D in individual sectors and technologies. But the subject remains full of controversy and unsettled issues. I believe that a work focussing as this one does on the R and D process in a concrete individual setting is valuable as a way of grounding and testing in the experience of actual cases and sectors what we think we know about the subject. Given the other concern underlying the book, we have here a fruitful opportunity to examine how R and D efforts and the technological conditions they create (or fail to create) grow out of and react on the broader issues of economic policy that R and D is intended to serve.

The R and D element in energy policy is a vast topic, of course, as we know from the huge literature on the subject that has already grown up in our own society. This book cannot pretend to be comprehensive in covering the topic, and I have deliberately selected a limited number of aspects of the issue for detailed investigation rather than trying systematically to cover everything. Several major considerations have governed the choice of topics to concentrate on: First I have tried to consider R and D areas that are most relevant to current energy policy issues. For example, nuclear power is so central in energy policy that it is absolutely necessary to deal with it. Unfortunately some things that are clearly important may not be researchable because of inadequate information in Soviet literature. Thus I have been able to deal only superficially with long-distance power transmission. Also, I lack the engineering expertise or the practical acquaintance with technological conditions in other countries to deal with some important issues, such as many aspects of underground coal mining, and have therefore tried to focus on questions or levels of generality in which conclusions may be obvious enough not to require that kind of expertise. Finally, since I have dealt with some of these issues of technology and R and D in the oil and gas industry in other works (*The Economics of Soviet Oil and Gas* and *Trends in the Soviet Oil and Gas Industry*), I have rather slighted the oil and gas sector here. The examination in Chapter 7 of technology transfer is based on cases drawn from that sector, however.

Given the goal of putting technological policy in its overall problem setting, Chapter 1 offers an overview of the Soviet energy situation and Soviet energy policy as background. R and D and innovation in the USSR take place in a different institutional setting from that in the United States, and, for those not familiar with that setting, Chapter 2 prefaces its description of the energy R and D establishment with an explanation of how energy R and D policy is made. Chapters 3 through 7 are essentially case studies of individual sectors and technologies. Chapter 8 attempts to draw together some implications for the two main issues that have motivated the book.

The research for this work was carried out under a grant from the National Science Foundation (NSF SOC 74–17609) and the Foundation's financial support is gratefully acknowledged. I would also like to express my appreciation to Stephen Able and Judith McKinney who provided valuable assistance in gathering material for the book. Much of the text of Chapters 2 and 3 originally appeared in two studies prepared for the RAND Corporation (Campbell 1978a and 1978b), and the section on gas pipelines in Chapter 6 was originally written as a study paper for the California Seminar on Arms Control and Foreign Policy. I am pleased to express my thanks for the permission of these organizations to use that material in this book.

SOVIET ENERGY TECHNOLOGIES

1

The Energy Sector and Energy
Policy in the Soviet Economy

Before proceeding to our main interest—the examination of Soviet energy technology and R and D efforts—it will be useful to have in mind a general picture of the energy situation in the Soviet Union and some of the main lines of energy policy it has evolved over the years. This chapter describes the Soviet energy resource base, some distinctive features of the structure of the supply and demand for energy in the Soviet economy, and the establishment of the overall framework of energy policy. The final section comments on the energy sector's importance as a claimant on national economic resources.

ENERGY RESOURCES

The nature of the Soviet energy resource base has important consequences for Soviet energy policy and for the direction of its efforts in energy R and D. The Soviet Union's energy resources are large enough that it can pose the trade question not as one of how much to import, but as one of how much to export. It is the only large industrial country in that position. But the economic characteristics of its energy resources are not especially attractive, and most of its internal energy choices and the direction of its energy R and D are conditioned by the location, quality, transportability, and other economic features of these energy resources. The best way to convey the situation is to describe briefly the major components of its energy resource base.

Coal *

Soviet authorities like to say that the USSR has over half the coal resources of the world. This is based on their estimate of 3,669.5 billion tons of "general-geological" reserves, down to 1,800 meters and in seams at least 0.3 meters thick for hard coal and 0.6 meters thick for lignite.

More important for current fuel planning are the records kept by the All-Union Geological Fund of explored† and commercially pro-

---

*The following description is based mostly on Mel'nikov, 1968b.

†This refers to reserves in the $A+B+C_1$ categories of the Soviet reserve calculation. The definition of these categories is rather detailed, but the concept is intended to describe reserves sufficiently well explored to justify production decisions.

1

ducible (*balansovye*) reserves, estimated at 237.2 billion tons on 1 January 1966. The important point is that explored and commercially usable resources of coal are adequate to support production for a couple of centuries even at a rate of production well above the current 750 million tons per year.

Economic evaluation of Soviet coal resources must take account of such economic characteristics as suitability for coking, heat and ash content, suitability for strip mining, cost of production, and location. The major features that should be kept in mind can be covered by several generalizations.

(1) The most intensively developed source—the Donets basin, attractive for the quality of its coal and its location near the sources of demand—has been exploited for a long time. Production must now penetrate deeper and resort to less economical portions of the basin. Production costs and incremental capital requirements for Donets coal are now very high.

(2) The cheapest sources to produce tend to be poorly located with respect to demand; the high quality coals of the Karaganda, Kuznetsk, and Ekibastuz basins are relatively cheap to produce, but must be hauled long distances to the markets they serve. The cheapest coal to produce in the USSR—in the Kansk-Achinsk basin in eastern Siberia—is a long way from the main regions of increasing energy demand.

(3) The best prospects for expanding coal output are in very large open-pit mines in western Siberia and Kazakhstan—the Kuznetsk, Ekibastuz, and Kansk-Achinsk basins. In addition to the common disadvantage of transport cost, Kansk-Achinsk coal presents difficulties in utilization because of low quality. This coal has a high moisture content (which complicates handling in winter), high ash content, low calorific value, and a tendency to spontaneous ignition. When dried, its high friability makes it difficult to handle.

Natural Gas and Condensate *

After a very late start, an extensive exploration effort in the sixties had raised natural gas reserves to about 25 trillion cubic meters in the $A + B + C_1$ categories. That is an amount several times the explored reserves in the United States. It is likely that explored reserves will continue to expand rapidly as exploration continues. In addition to reserves in the $A + B + C_1$ categories, the Russians estimate large

---

*This section is based mostly on data in Vasil'ev, 1975, which gives comprehensive tabulations and detailed descriptions of all the fields and regions.

holdings of "probable" reserves (their categories: C_2, D_1, and D_2). The most recent estimates of these probable reserves seem to be for 1 January 1971, when they were estimated as 76 trillion cubic feet, down to 5,000 meters, with another 10 trillion cubic feet predicted in the interval from 5,000 to 7,000 meters. At current rates of exploration large amounts of these probable reserves will be transferred to the $A + B + C_1$ categories. Soviet data for gas reserves include associated gas in gas caps, but not dissolved in oil. Furthermore, it is not certain whether they are comprehensive for offshore gas.

As with coal, the economic characteristics of these gas resources diminish their attractiveness. In both the explored and probable categories, most reserves are in the eastern part of the country and a considerable share is in the far north. Of the 22.6 trillion cubic meters of $A + B + C_1$ reserves on 1 January 1974, 13.9 trillion cubic meters were in Tiumen' *oblast'* in western Siberia, while the rest of Siberia accounted for another one trillion cubic meters. Central Asia and Kazakhstan accounted for 3.9 trillion cubic meters, leaving only 3.4 trillion cubic meters in the European part of the USSR. The only really large source in the European USSR was the Orenburg field, though the Komi ASSR also had considerable reserves, relatively poorly located. This locational pattern has two important consequences: conditions of climate and terrain are adverse for the exploration and production of gas fields and for the construction and operation of the associated pipelines. In addition, the transmission distance to the major centers of demand is very great. This is an especially adverse circumstance for gas, since its low density makes it expensive to transport.

The Russians do not publish systematic figures on condensate reserves,* but they are no doubt very large. Thus far, little attention has been given to the capture of condensate. Most condensate reservoirs are currently produced without repressuring, which leads to losses of condensate as reservoir pressures drop, causing condensation within the reservoir.

Oil

Soviet sources do not generally release absolute data on oil reserves, since this is prohibited by the State Secrets Act, and we must discuss Soviet oil potential in terms of rather general considerations.

---

*Condensate is formed by hydrocarbons of the C_6 type (and higher), which at reservoir temperatures are dissolved in the gaseous phase, but which condense at the lower temperature and pressure at the surface.

The USSR has very large volumes of sedimentary cover promising for oil accumulation. Large areas of these sediments have been explored only lightly because of the delayed development of the oil industry in the USSR. Once the drive began to expand oil output, relatively modest outlays on exploration disclosed large reserves. Though the growth of reserves has both led and lagged the growth of output at various stages, in recent years the ratio of reserves to output has fallen. But that is probably more an indication of planning errors than of resource exhaustion. The biggest prospects for additional reserves seem to be in offshore areas, in the arctic regions of Siberia, and at depths below those customarily explored so far (below about 3,000 meters). To keep output growing, even to keep it stable, will require meeting some stiff technological demands posed by the new environments.

Oil development has so far followed the geographic pattern common to all major energy resources—early development in European regions with subsequent depletion requiring a shift to Siberian and Central Asian sources. The locational pattern for oil resources still to be found (if technological obstacles can be overcome) should be more favorable than for those currently being produced; the new resources are closer to the big market areas of the European USSR. Specifically, there are thought to be considerable additional resources at depths of about 3,000 meters in the Caspian depression and in offshore areas adjoining the European part of the USSR.

Hydroelectric Power *

The USSR has large resources of hydroelectric power, though most of it is still undeveloped. Translated into fossil fuel equivalents at the heat rate attained in thermal stations of the utility network, hydroelectric power output now contributes a little over 3 percent of all primary energy output.

As with other energy sources, hydroelectric potential is heavily skewed toward the eastern part of the country. An inventory and economic evaluation of hydroelectric potential conducted during 1958–1965 indicated the economically developable potential under present technological conditions at about 1,095 billion kilowatt hours of annual output—i.e., an output roughly equal to total electric power output in the mid-seventies. Since the definition of developable was that "which in the light of contemporary views on the development of

*This section is based essentially on Neporozhnyi (1970).

electric power and in the light of economic evaluations can be used in the near or distant future," it would appear to assume considerable technical change and location accommodation (Neporozhnyi, 1970, p. 236).

At the beginning of the seventies, 43 percent of the economic potential in the West (201 BKWH) was in use or being developed, while in the East the corresponding share was only about 16 percent. Thus locational considerations loom large in controlling how fully potential hydropower resources can be used. An attempt has been made to take advantage of Eastern hydro resources by locating such energy intensive industries as aluminum reduction near the sources. But this can make only a slight difference, and significant utilization of Siberian and Central Asian potential over the next couple of decades on a large scale depends on developing new long-distance transmission technologies. There is already some movement of Siberian-generated hydropower through the interconnections of the Siberian grid with the European grid, but the amount of this flow is very small.

Hydroelectric resources have always stimulated grandiose dreams in Soviet planners, and some projects for solving the locational problem by interbasin transfers of water have long attracted attention. Most of these are concerned with irrigation water rather than with hydro potential, but one project, the diversion of water from the northern rivers to the Volga basin, would also have significant effects in increasing the power potential of the cascade of dams already in existence on the Volga. The Soviet planners seem to have serious intentions of making this diversion at some time.

Minor Fuels: Peat, Shale, Firewood

The Soviet energy balance differs from that in most developed countries by using appreciable amounts of minor fuels. The resources are large, and their location has made them attractive supplements to standard resources in some areas.

Peat occurs very widely in the USSR, but has been exploited mostly in the West, Northwest, and Center regions, which are deficient in other fuels. More effort has been devoted in the Soviet Union than in other countries to make peat a significant fuel resource. A number of distinctive technologies have been developed for extracting, transporting, and processing peat into briquettes and other concentrated fuels, and equipment to burn it on a fairly large scale in electric power stations has been developed. The problem, of course, is to handle it in large volumes, and most of the power stations burning peat have been rather small. The largest peat-burning station is the Shatura station (732 MW capacity) and there are a number now being constructed of 600 MW capacities (Popov, 1974, p. 708). In 1940, 20 percent of all electric power output was produced in peat-burning stations (Mel'nikov, 1968b, p. 569). But the revolutions in the fuel balance and in transport technology for gas and oil have made peat uneconomic as an energy fuel even in regions poor in other energy resources, and its output and use have remained fairly stable at 50–60 MT per year in the last two decades.

Oil shale has played a minor but not insignificant role in the Soviet fuel balance for a long time, again primarily under the protection of distance and the absence of better local alternatives. Its contribution to fuel output is only about a third of that made by peat. Oil shale resources are located in two main areas, the Volga region and the Baltic coast.

Oil shale is burned directly in boilers and furnaces and is also retorted to recover the kerogen or to produce various products by destructive distillation. The potential for using oil shale as a large-scale source of liquid hydrocarbons seems to be governed by much the same considerations as in the United States. The Soviet deposits are better located with respect to water, but the problems of environmental damage from large-scale operations in the USSR are serious. The Soviet oil shale industry has not been notably successful in reclaiming land damaged by mining or in disposal of processed shale. The general current attitude toward oil shale seems to be that if its use is to expand, it should be as a source of materials for the chemical industry (Mel'nikov, 1968b, pp. 621–627).

Firewood has remained a significant fuel both for many small scale local industries close to the lumbering areas and for households. It is an expensive alternative, of course, and the persistence of firewood use is based on the traditional unconcern of the system with modernization evenly across all sectors as well as its general de-emphasis of consumption compared to other end uses of GNP; this has left households, especially rural households, to shift for themselves.

Nuclear Resources

Little is known about Soviet resources of fissionable materials. There seems to be a general consensus that the USSR obtains much of its uranium from Eastern Europe. I have never seen any explicit Soviet discussion that would illuminate how fully the USSR has estimated or explored its own domestic uranium and thorium resources or what kind of supply curve is estimated. This is a subject intimately connected with nuclear power policy, and further discussion will be deferred to Chapter 5.

The USSR also has extensive resources of other novel primary sources, such as solar, geothermal, and tidal energy. These play virtually no role in the current energy economy and can be ignored in an analysis of the main issues of current energy policy. They are considered important for certain regions and for the future; the R and D efforts to develop technologies to utilize them will be examined in Chapter 6.

In sum, the USSR is one of the world's "have nations" in relation to energy supplies. The Soviet situation becomes less favorable when one takes account of the economic characteristics of these resources. They are not well-located, many are expensive to produce, the quality of many sources is low. In many cases, novel technologies will be required to produce and utilize them extensively. Many Soviet energy sources can be utilized only at the risk of considerable environmental damage, and much damage has already been done through loss of land to reservoirs and strip mines, pollution of rivers with petroleum wastes, and air pollution from burning low quality coals. The best characterization is probably that although the USSR has abundant energy supplies, it is not a country of cheap energy.

The locational factor is one of the most important forces constraining and conditioning energy policy and defining energy R and D tasks. It underlies the current emphasis on atomic power, for example. Despite the abundance and low cost of resources in the East, transport costs give a strong economic advantage to nuclear power west of the Urals. The uneven distribution of water resources (important both as a power resource and for cooling thermal stations) shape Soviet energy policy in distinctive ways, such as the long-standing fascination with big projects for the interbasin diversion of water and the great interest derived therefrom in peaceful nuclear explosives. The regional problem has a heavy influence on the direction of Soviet innovative effort—it provides a special motivation for innovation in long-distance transmission of electric power and for novel transport modes for other kinds of fuel.

Despite the abundance of most of the traditional energy resources, the USSR now finds it necessary to use much more sophisticated technologies in processing and utilization than it has in the recent past to compensate for the low quality of many energy sources and to enhance energy efficiency in view of rising costs. Examples include complicated cycles in thermal power generation, elaborate processing of coal to ameliorate the environmental impact of its use and to make it more transportable, and more sophisticated refining of petroleum to produce products for high-grade uses.

ENERGY PRODUCTION
AND CONSUMPTION

Energy policy and energy R and D are shaped by the peculiarities of demand in any society, and it will be useful to have in mind some of the major elements of the energy supply and demand situation in the Soviet economy. Information on the production, transformation, and utilization of energy in the USSR is not available in Soviet sources in systematic and complete form. Anyone who wants to study this subject must start by developing his own overall energy balances for the Soviet Union. This is a complicated task, and I intend to sketch here only some of the broadest features of the situation important for understanding Soviet energy policy. The reader who is interested in the more detailed analysis and documentation for these generalizations is referred to a much fuller account in *Soviet Energy Balances* (Campbell, 1978b), on which the following section is largely based.

Growth and Composition of Supply

Table 1-1 shows the history of primary energy output by major source over the whole period of Soviet industrialization and the disposition of that output between domestic consumption and export. During the first 30 years of the Soviet industrialization drive, energy production grew considerably more rapidly than national output. While the growth of primary energy output between 1928 and 1958 averaged about 8.7 percent per year, Abram Bergson has estimated the average annual rate of growth of Soviet national product at a little over 6 percent (Bergson and Kuznets, 1963, p. 6).

Energy growth is measured here in terms of heat content; in view of changing composition and quality, energy output in value terms probably grew still faster. Energy production's outpacing of total output seems to have been a persistent phenomenon, slightly perturbed in some periods by the changing relationship of trade to production. Overall, since the share of exports fell during this period, the growth of energy *consumption* at 8.8 percent per year was still higher in relation to GNP growth than was energy *production*.

A significant break in the energy/GNP relationship came toward the end of the fifties when a new energy policy (see below) made possible a number of fuel economizing shifts, and the growth of energy consumption dropped to about the same rate of growth as GNP. During the sixties energy output grew at 5.3 percent per year, while GNP grew at that same rate (CIA, 1977). In the seventies, Soviet year-to-year GNP growth has been erratic, but has dropped compared to the sixties, averaging about 4 percent. The rate of growth of energy consumption has remained at about 5 percent in the seventies, however, so that the elasticity of energy consumption with respect to GNP growth has again moved well above 1. We might add in this connection that international comparisons of energy consumption have generally found that the USSR is a fairly energy-intensive case—i.e., it has relatively high energy consumption per dollar's worth of GNP (see Darmstadter, 1971, pp. 32–40). The high elasticity with respect to GNP growth and the high energy intensity of GNP are probably explainable in large part by the same factors: emphasis on energy-intensive outputs and relatively low efficiency in the use of energy.

The composition of energy output has undergone several shifts associated with a marked periodization of energy policy. Two early policy goals were the replacement of wood by fossil sources of power and an effort to avoid strain on the transport system. In connection with the latter goal, strenuous efforts were made to develop local fuel sources in every region. These sources were often of low quality, such as peat, lignite, and oil shale in the Northwest. Because of an insufficient effort to discover new oil resources outside the traditional regions, the growth of oil output fell behind that of solid fuel.

The Second World War forced an accelerated effort to develop oil outside the traditional major oil producing region in the Caucasus and TransCaucasus, specifically in the Volga-Ural region where large resources were found. After the war oil output began to rise fairly rapidly, increasing its share from 18 percent of all fossil fuel output in 1945 to 23 percent by 1955. But then a radical reorientation of the fuel balance took place under new policies introduced as part of the Seven Year Plan (1959–1965), with both oil and gas commencing to grow much faster than other kinds of energy output. Their share in fossil fuel output rose steadily and by 1975 was two-thirds of the total. The best summary expression of this shift in emphasis is found in the

TABLE 1-1. Soviet Fuel and Energy Production, Trade, and Apparent Consumption

(million tons of standard fuel[a])

Year	Mineral Fuels						Fire-wood	Hydro[b] power	Nuclear[b]	Total energy	Net[c] trade	Apparent Consumption
	Coal	Oil	Gas	Peat	Shale	Total						
1928	29.8	16.6	0.4	2.1	n.a.	48.9	5.7	0.2	—	54.8	−3.9	50.9
1940	140.5	44.5	4.4	18.6	0.6	203.6	34.1	3.3	—	241.0	+2.3	243.3
1950	205.7	54.2	7.3	14.8	1.3	283.3	27.9	7.5	—	318.7	+10.0	328.7
1958	362.1	161.9	33.9	21.1	4.5	583.5	32.9	22.5	—	638.9	−27.4	611.5
1960	373.1	211.4	54.4	20.4	4.8	664.1	28.7	23.8	—	716.6	−51.1	665.5
1965	412.5	346.4	149.8	17.0	7.4	933.1	33.5	33.8	.6[d]	1,000.4	−101.5	898.9
1970	432.7	502.5	233.5	17.7	8.8	1,195.2	26.6	45.6	1.3[d]	1,267.4	−153.6	1,113.8
1975	490.4	701.8	345.7	16.9	11.7	1,566.5	23.8	42.8	6.9	1,640.0	−201.4	1,438.6
1977	486.0	780.5	410.0	14.0	11.4	1,701.9	24.6	49.1	11.4	1,787.0		
1980 Plan	565.2	915.0	519.8	(15)	(13)	2,028.0	(27)	67.0	27.2	2,149.2		

SOURCES AND NOTES: Based on Campbell (1976), supplemented with current statistical handbooks.

Figures in parentheses are estimated.

aStandard fuel is the common denominator used in Soviet energy accounting. One ton of standard fuel = 7×10^9 calories.

bConverted at the fuel rate for thermal central stations of the corresponding year.

cNet exports shown as − and net import as +.

dNot included in totals.

fact that between 1958 and 1975, oil and gas accounted for 85 percent of the increment in primary energy production.

The availability of very large increments of hydrocarbon fuel, at relatively low cost, both sustained and in part prompted a change in general economic policy toward a more open stance with respect to the rest of the world. Under this policy, large amounts of technically advanced capital goods imports were made possible mainly by earnings from energy exports. Most of the increment in energy exports has been accounted for by oil. In the mid-seventies, 40 to 50 percent of Soviet hard currency earnings came from oil exports. Coal, coke, and electric power have also been exported, but in 1975, they amounted to only 12 percent of energy exports by energy content, slightly more in terms of value. The fact that energy exports are skewed towards oil means that the structure of consumption is slightly different from the structure of production.

The post-1958 experience of the Soviet Union with regard to fuel composition obviously parallels in many respects trends that had taken place much earlier in the United States and Western Europe; at the same time there are some differences. Despite the shift to hydrocarbons, the USSR has remained more dependent on solid fuels than either the United States or Western Europe. In 1975 solid fuels, which constituted only 19.1 percent of total primary energy consumption in the United States and 21.5 percent in Western Europe, were 36.8 percent in the USSR. The large role for solid fuel, moreover, involves significant resort to *low-grade* fuels—of the 36.8 percent of total energy consumption accounted for by solid fuel in 1975, 11.2 percent is peat, shale, firewood, and lignite, all of which entail high costs in their production, transport, and utilization. The substitution of hydrocarbon for solid fuels has probably gone about as far as it can in terms of shares, and *within* the solid fuel category, dependence on low-grade coals will increase. One of the most pressing tasks for Soviet energy R and D is to improve or develop technologies for handling low-grade solid fuels.

The USSR made a slow start in developing nuclear power and for most of the years shown in Table 1-1 its contribution to total primary energy output is so small that it could be ignored. The USSR still obtains a smaller share of its total energy supply from this source than either the United States or Western Europe. In 1975 nuclear sources provided 0.5 percent of Soviet consumption of primary energy, 2.3 in

Western Europe and 2.5 in the United States. The slowness of nuclear development is partly explained by the relative abundance of other fuels available to the USSR; but, as explained earlier, despite having *abundant* energy resources, the USSR is not a country that enjoys *cheap* energy supplies, and there has been a strong rationale for nuclear power for some time, particularly on a regional basis. Soviet energy planners were just slow in making the decision to push nuclear power, and the prospect is that the share of nuclear power in Soviet energy supply will catch up and even surpass that in other areas within a few years.

Somewhat in contrast to what one might expect, the USSR has a low share of total supply accounted for by hydroelectric power. Despite a long-standing bias in favor of hydroelectric power, its share has been appreciably below that in Western Europe and even somewhat below that in the United States.

Energy Consumption by Sector

Consumption of energy by sector is shown in Table 1-2. Several facts stand out in this comparison with the United States and Western Europe.

Losses and internal consumption within the energy sector constitute a higher share of output than in either the United States or Western Europe. Some of this internal consumption is caused by conditions over which Soviet planners have little control. These include the long distance (and hence, high energy cost) for transporting oil and natural gas, the large share of coal converted to coke, and consumption losses associated with the use of low-grade fuels. Still, these losses probably merit more attention in fuel policy than they get.

The most distinctive feature of the Soviet use pattern is the relative unimportance of transport both in absolute terms and with respect to its share in the total. In 1975, the USSR used only about one-seventh as much energy for transport as the United States, mostly because the Soviets have a much smaller stock of automotive vehicles, especially private passenger automobiles. (It may be that the small share is somewhat exaggerated, reflecting possible failure in these balances to get all automotive transport under the transport heading.)

I might mention, however, that despite the even lesser importance of automotive transport in 1960, the transport sector's share of gross domestic energy use in that year was nearly as high in the USSR as in Western Europe. This high consumption was due to the predominance

TABLE 1-2. Energy Consumption by Using Sector, 1975

	Million tons of standard fuel			Percent of total[b]		
	USSR	WE	USA	USSR	WE	USA
Total domestic energy use	1482[c]	1605	2412	100	100	100
losses and internal consumption	186	94	144	12.6	5.9	6.0
Nonenergy uses	64	54	58	4.3	3.4	2.4
Electric power	487	457	713	32.9	28.5	29.6
Industry	407	392	511	27.5	24.4	21.2
Transportation	85	234	583	5.7	14.6	24.1
Other[a]	253	374	403	17.1	23.3	16.7

[a]Residential and commercial, construction, public uses, agriculture.

[b]In some cases, components do not add to total because of rounding.

[c]Differs slightly from the figure shown in Table 1-1, since this figure includes in gas output, gas flared (21.5MT of standard fuel), and noncommercial energy in the form of firewood and peat gathered by the population (20 MT of standard fuel).

of railway steam traction in Soviet transport as a whole and the fact that steam traction is very inefficient with respect to energy inputs. The decline in the transport share has resulted from the shift to diesel locomotives, which do a given amount of work with a much smaller energy expenditure.

The share of total energy supply converted to electric power is higher in the USSR than in either the United States or Western Europe. The difference has usually been a few percentage points, although my balances may understate somewhat the amount of fuel resources consumed by electric power stations in the USSR. I have depended mostly on Soviet statements regarding the heat rates achieved, which I suspect show somewhat exaggerated performance. (More will be said about this in Chapter 3.)

In terms of shares, the high share for electric power is largely just the opposite side of the low share for transport in the USSR. In absolute terms, the USSR converts much less energy to electric power than does the United States, and slightly more than Western Europe.

The share of energy consumed for industry is higher in the USSR than in either the United States or Western Europe and lower for the "other" sector, comprising residential and commercial, construction, public uses, and agriculture. In the USSR in 1975, industry used 1.6

times as much energy as the "other" sector, whereas in Western Europe it used only 1.05 times as much, and in the United States only 1.3 times as much. The comparison is even more striking if we look at the ratio of consumption in industry to consumption in a more restrictive concept of household use. We can separate out "residential" use for both Western Europe and the United States in 1975, and a "household and municipal" category within the Soviet total for "other" with the resulting ratios as follows:

USSR	2.6
Western Europe	1.2
United States	1.5

These ratios are probably a reflection of the peculiarities of Soviet final demand structure, in which household consumption is depressed relative to investment and military expenditures. The low level of per capita consumption means low household demand for energy, and industry gets the difference.

In the full fuel and energy balances for the USSR, the shares indicated for agriculture and for construction are significant. These are uses that are so small in Western Europe and the United States that they do not usually even receive mention, much less routine segregation. The source for Western Europe in 1975 does show agriculture as a separate sector (consuming 25.2 million tons of standard fuel), and the U. S. Department of Agriculture has estimated energy use in U. S. farm production in 1974 as 46.8 million tons of standard fuel (without electricity, the amount would be slightly less—FEA, 1977, vol. 1, p. 2) compared with 69.8 million tons of standard fuel for the USSR. Since Soviet agricultural output was probably no more than 80 to 85 percent of U. S. agricultural output in the 1970s (U. S. Department of Commerce, 1972, p. 43), there is a suggestion here that fuel and energy are used very inefficiently in Soviet agriculture.

Another way to look at energy consumption is to treat electric power production as an intermediate transformation rather than as a final use. When the energy consumed in power stations, net of conversion losses, is reallocated to the final consuming sectors, the results for 1975 are as shown in Table 1-3. I have simplified the table still further by eliminating losses and internal consumption and nonenergy uses, and looking only at the relative importance of the three main final demand sectors. Looked at in this way, the predominance of industrial use in

TABLE 1-3. Fuel and Power Consumption by Major Sectors, 1975

	Million tons of standard fuel*			Percent of total*		
	USSR	WE	USA	USSR	WE	USA
Total consumption	983	1152	1724	100	100	100
Industry	580	468	601	59.0	40.6	34.9
Transportation	94	241	583	9.5	20.9	33.8
Other	310	443	540	31.5	38.4	31.3

*In some cases, components do not add to total because of rounding.

the USSR is enhanced still further. The heat and power outputs of Soviet electric power stations go even more heavily to industry than do direct fuel inputs.

The USSR consumes nearly as much energy in industry (580 million tons of standard fuel in 1975) as does the United States (601 million tons of standard fuel). This is remarkable considering that Soviet industrial output is appreciably smaller than U. S. output. Determining the relative size of such economic aggregates as industrial output is an ambiguous business, but it is usually said that Soviet industrial output is probably no more than three-fourths of the industrial output in the United States. The Soviet Central Statistical Administration says that Soviet industrial output was about 80 percent of the U.S. level in 1975. This heavy use in industry may be attributable in part to a more energy-intensive industrial structure than in the United States. This is a more complex matter than it sounds, and I will not go into it here. But it is not unambiguously clear that the Soviet industrial structure is a great deal more energy-intensive than the U.S. structure. Even if differences in industrial structure do play some role here, the difference in energy use per unit of output is so large that it seems highly likely that energy use in Soviet industry is very inefficient.

Having moved from treating electric power as a final consumer to treating it as a transformation technology, the question arises of the efficiency with which the electric power sector transforms energy. As will be explained in Chapter 3, the USSR has relied heavily on cogeneration as a technology for its electric power sector, and this has an important effect on conversion efficiency. When we compare the ratio of energy in electric power output to the energy in the fuel burned in electric power stations in the USSR with that in the United States and

Western Europe (see the tabulation below), the Soviet conversion ef-

USSR	0.2447
WE	0.3592
USA	0.3815

ficiency looks very low (these data refer to 1975). But the USSR captures a large amount of by-product heat from the generation of power and uses it for space heating and industrial process heat. When this output is accounted for, the Soviet conversion efficiency rises from the 0.2447 shown in the tabulation to 0.5141, significantly *higher* than for the United States and Western Europe, whose conversion efficiencies would be modified only negligibly by this correction.

The Soviet experience underlines in a dramatic way the energy-saving potential of co-generation, and should heighten our own interest in this method as a possible energy conservation measure. The Soviet Union, having already exploited that conservation tactic, must direct its search for fuel savings in electric power generation along other lines.

Some of the most distinctive differences in Soviet energy use patterns are revealed when we look simultaneously at type-of-fuel and sector-of-use. We can best get at this by a USSR/Western Europe/United States comparison of the amounts in particular cells of a source/use matrix. For example, the following data for 1975 show the consumption of petroleum products (in million tons of standard fuel) in the generation of electric power, and in transportation:

	USSR	WE	U.S.
Transport	75.9	234	583
Electric power	137	101	112
Ratio of transportation/electric power	0.55	2.3	5.2

The tenfold difference in the ratio between the United States and the USSR is a compound of all the sharpest differences between the supply and demand structures, especially the differential importance of transport, and the differing importance of petroleum compared with other energy sources. Similar disparities show up in the uses to which gas is put, as indicated below (for 1975, in million tons of standard fuel):

	USSR	WE	U.S.
Electric power	87.7	43.2	104.4
Industry	130.8	86.4	316.8
Other plus transport	54	79.2	223.2

Given the low share of household demand in its market structure, the USSR has not needed to allocate much gas to the "other" sector, and has absorbed the rapid growth in gas output in part by using it more heavily in the electric power sector than does the United States or Western Europe.

These differences in Soviet energy consumption structure suggest an important conclusion for energy policy: to the extent that the Russians want to employ energy conservation as a strategy, their attention should probably be directed along somewhat different lines than ours. Given the small size of the household and commercial sector in the total and the small size of the transport sector, the measures that have attracted so much attention in U. S. energy conservation efforts—economizing on space heating, improving automobile efficiency, discouraging the manufacture of high-horsepower automobiles, and raising the efficiency of household appliances—are not very promising for the USSR. Rather, Soviet efforts need to be directed primarily at measures to save energy in industrial processes. I suspect that because of systemic differences, energy conservation in industry may be more difficult to achieve in the USSR than in market economies. In the United States, the stimulus of higher energy costs leads businessmen themselves to seek out many ways to save energy, even in the absence of such government measures as taxes, subsidies, and special R and D efforts. In the Soviet Union, the system is much less likely to show this kind of automatic response; Soviet managers are much less likely than capitalist firms to respond to cost pressures, and it is inherently difficult to attain these savings by campaigns from above because the potentials are so scattered and varied. The Soviet rationing system probably leads to wastage of fuels in the nonhousehold market in a way that has often been noted by economists both with respect to rationing in other countries and in the experience of the Soviet Union. Customers pad requests in the expectation that they will be cut, which leads the rationers to arbitrary cuts, and so on in a vicious circle.

Until recently, one of the interesting characteristics of the Soviet literature on energy policy has been its limited interest in conservation, and Soviet energy officials are only now beginning to speak seriously of such measures as price policy to encourage conservation. With a few exceptions this is also true of the literature on energy R and D, and in dealing mostly with supply-side technologies, rather than with energy utilization technologies, this book accurately reflects Soviet preoccupations. The main exceptions are in the discussion of electric power, where there has been a strong focus on the heat rate as a technological indicator toward which innovation is directed.

ORGANIZATION AND MANAGEMENT
OF THE ENERGY SECTOR

The Soviet economy is a planned economy, and the multifarious decisions in which an overall energy policy is expressed—choice between alternative energy sources, regional and locational decisions, choice of technologies, the allocation of resources to R and D and the direction of R and D efforts—derive from the general planning process that determines all allocation decisions in the Soviet economy. It is neither possible nor pertinent to try to describe that setting in detail here, and the reader unfamiliar with it is referred to standard treatments of the institutions, processes, and strategies that distinguish the process of resource allocation in the Soviet economy. (See, for example, Bergson, 1964, Campbell, 1974, or Gregory and Stuart, 1974.) It may be useful, however, to describe how the energy sector is organized and to explain the approach that has been used to establish the grosser lineaments of energy policy.

Organizational Structure

The distinctive feature of Soviet energy management is that it is effected through a hierarchical administrative structure. The management of the energy sector is primarily in the hands of several ministries: The Ministry of Electric Power and Electrification (Minenergo), The Ministry of the Coal Industry (Minugol'), The Ministry of Oil Extraction (Minneft'), The Ministry of the Gas Industry (Mingaz), The Ministry of Oil Refining and Petrochemicals (Minneftekhim), The Ministry of Geology (Mingeo). These main branch ministries are also in charge of unconventional energy sources—Minenergo has jurisdiction over tidal and nuclear power and shares responsibility for geothermal energy with Mingaz. Minneft' has jurisdiction over extraction of oil shale.

The technological level of each of these industries is determined to a considerable extent by the capital goods available to it. The electric power industry depends on the Ministry of Electric Equipment (Minelektrotekhprom—generators, switch gear and transforming equipment), the Ministry of Power Equipment (Minenergomash—turbines and boilers), and the Ministry of Control Equipment (Minpribor—

control devices) to increase its output and improve its technology. The oil and gas industry depends on the drilling equipment it gets from the Ministry of Petroleum Machine Building (Minneftemash) and the Ministry of Heavy and Transport Machine Building (Mintiazhmash) and on the pipe it gets from the Ministry of Ferrous Metals (Minchermet).

Taken together, these ministries control most of the research and development resources available for developing and improving energy technology. Each has a large network of R and D organizations under its control. Some research and development activities for energy, however, are under the control of two organizations without responsibilities for energy production. The State Committee for the Peaceful Use of Atomic Energy is essentially an R and D ministry for nuclear technology and the Academy of Sciences controls numerous organizations doing basic research on energy. This energy R and D establishment and the planning of its activity will be more fully described in the following chapter.

A striking aspect of the administrative structure for the energy economy is that there is no energy czar—no energy commissar. There is no administrative node above the ministries we have mentioned with responsibility for energy as a whole. Executive control of these subdivisions of the energy sector is exercised only by the Council of Ministers as one aspect of its larger task of overseeing the whole economy. Many of the information threads come together in the Gosplan, which has a section on fuel and energy, but Gosplan has no executive power to make energy policy or decisions on its own.

There are two important consequences of this structure for energy policy. First, it is a polycentric system in which many of the issues of energy policy get settled through bureaucratic struggles and the political processes of an oligarchic power system. It is only in these terms that we can interpret outcomes on such important issues as the relative priority of nuclear versus conventional sources or the trade off between exporting energy resources to hard-currency trade partners versus other socialist countries. Unfortunately the working of these processes is often opaque, and an understanding of them is more likely to come via the skills of Kremlinology than those of economic analysis.

Second, this system is highly vulnerable to the administrative disease known as "suboptimization," in which individual units, by trying to improve their performance according to assigned criteria, keep the system as a whole away from a global optimum. In the Soviet system, ministries are powerful organizations with a strong proclivity to defend themselves against control from above and from outside pressures. The coal-mining ministry, for example, optimizes locally in ways that create great burdens for other sectors of the energy economy and push global system performance below what is possible. The coal miners seek to reduce production costs by shifting to open-pit sources and by mechanization. Unfortunately, this raises the ash and rock content of coal, subjecting the users to extra costs that often exceed the savings to the coal industry. The railroads are also forced to spend huge amounts of resources on carrying this useless ballast. It should be the function of some higher level organ to correct this, but the instruments at the disposal of the top level planners are inadequate. They lack the information to calculate the optimal cleaning level. Moreover, their incentive levers are insufficiently subtle to make the coal industry carry the cheapening effect of open-pit and mechanized mining just to the point at which, optimally offset by adequate beneficiation, coal quality is optimized in the sense of making production, transport, and using costs a minimum.

One of the most interesting features of the organization of the sector is the constantly shifting pattern of jurisdictions, intended to make the administrative structure more nearly isomorphic with production interrelationships. Both the outer border, where these ministries touch on other sectors of the economy, and the jurisdictional dividing lines within the sector are constantly being changed. Minneft' and Mingeo are in a constant struggle over the territory within which each will have authority to conduct exploration. The processing of by-product gas has frequently shifted back and forth between Minneft' (from whose wells it comes) and Mingaz (through whose pipelines it will be delivered to customers). Minugol' has recently won a struggle to take over from Mintiazhmash the plants that produce coal-mining machinery. When those plans were under the control of Mintiazhmash, they (and their supervising ministry) were not especially interested in serving the needs of their coal-mining clients. In the Soviet economy the resolution of such conflicts cannot be found by some kind of lateral negotiation and bargaining, but must be pursued through vertical communication channels. The paradoxical result is that, though planning might be thought to have precisely the virtue of eliminating such conflicts and ensuring the attainment of equimarginal conditions in all such situations, it is probably much less able to do so overall than is the market system.

Similarly ironic is the fact that, contrary to the connotation of a long-term perspective usually associated with planning, the Soviet administrative system in fact has a relatively short planning horizon. This weakness weighs especially heavily on energy management because of the long production cycles. The current difficulties in Soviet oil production probably owe a great deal to this kind of bias. Under strong pressure to meet current output goals, the industry responds by shifting resources from exploration to production, overdrilling known reservoirs, injecting water at rates above what would optimize the rate of recovery from the reservoir, and so on. All of these expedients probably cost more in terms of future sacrifices (even properly discounted) than they contribute to some concept of current welfare.

Forecasting and Modeling

In recent years the major method used to impose some sector-wide perspective on decisions about fuel policy has been a modeling effort aimed at optimizing a number of major fuel policy variables. In the process, some progress has been made toward creating an overall framework that would impose a resolution of some of these priority conflicts that is objective and rational from a national-economic view.

There is extensive Soviet literature on this modeling and forecasting work, much of which is methodological and schematic, describing basic concepts of forecasting and modeling and how it might be done, rather than how these schematic approaches are applied in practice.* Also, much of this work is at the level of subsectors rather than the energy sector as a whole, and a great deal of it is also concerned with still smaller units such as a region or an enterprise. But our major concern here is with the scenario-building and modeling work done from a national perspective by a group of research institutions in connection with the effort to project a fifteen year plan (1976–1990) and draw up the 10th Five Year Plan (1976–1980) for energy. The institutions include the Siberian Energy Institute of the Siberian Division of the Academy of Sciences, one of the major project-making organizations in the electric power ministry (Elektroset'proekt), the Main Computer Center of the Gosplan, and the major R and D in-

---

*Some sources, however, give much more informative descriptions of the form the actual forecasts take. For example, a book on forecasting the development of the coal industry to 1990 and 2000 lists the 27 volumes and 54 appendices that are to be produced under the forecasting assignment, together with a general description of the topics to be included in each and the institutions that are to produce them (Stugarev, 1976, pp. 64ff.).

stitute in the oil industry (VNIINP). The following description of their work, based on an article in *Planovoe khoziaistvo* (1975:2, pp. 29–37), provides some interesting insights into how the presently held strategic views on energy policy were generated. Although that article is somewhat elliptical, it is fairly easy to fill in the gaps because other sources provide much fuller descriptions of the kind of energy-modeling work these organizations have been doing.*

The first step is to outline a number of energy scenarios by putting together various combinations of a few projected variables. The authors are not specific as to the year, but most of the references in various sources imply that the target year is 1990. One of the variables is a forecast of energy demand, but with variations in the range of 10–15 percent. It is not at all clear how the demand forecast is made; one of the most obvious gaps in the whole Soviet literature is the absence of informative discussions of energy demand forecasting. The literature is full of abstract discussions of correlation methods, international comparisons, Delphi approaches, and other exotic techniques, but what one authoritative work on electric power planning says is probably about right: "At the present time the method of direct calculation via output targets and input norms is the basic one used in forecasting and planning consumption of electric power and other kinds of energy in the USSR" (Beschinskii and Kogan, 1976, p. 125).

Ceilings are specified for output of various primary sources, broken down by regions. These are apparently based on reserve estimates in some cases or, in other cases, on estimates of possible capacities (say for nuclear stations). For the particular exercise described in the article, some sources (specifically Siberian oil and gas) were treated as unconstrained. Using estimates for both investment and operating cost for these different sources, and for transport costs, the researchers set up a linear programming model designed to minimize the total cost of meeting the forecast demand. It is not clear in the description how much regional detail is specified in the demand forecast, to what extent the fuel and energy requirements are distinguished by type-of-use, and to what extent these demands are treated as permitting inter-fuel substitution. But the general energy planning models developed in these institutes depend heavily on these distinctions, since one of

*The most complete of these broader surveys is Makarov and Melent'ev, 1973. It has a bibliography of 139 modeling studies completed earlier. A summary and interpretation of the actual magnitudes for some important energy variables generated in these forecasts is available in CIA, 1975.

their objectives is to generate shadow prices differentiated by region and energy source, to be used in other planning calculations.*

The *Planovoe khoziaistvo* article indicates that about 80 scenarios were developed embodying various combinations of assumptions as to demand and output constraints. For each of these scenarios an optimizing calculation is made, for which the objective function is to minimize *privedennye zatraty* (outlays on current inputs plus an interest charge for the capital stocks involved in meeting the specified demands) in the terminal year of the forecast period. The solution in each case is an output pattern for the different main primary sources, a cost total, and (depending on how much regional and end-use detail has been used) a more or less detailed pattern of geographical flows and end-use allocations. There will also be a set of shadow prices from the dual of the linear programming problem described. For any given level of output, the plan with the lowest cost is the optimal plan.

The optimal plan derived from these calculations implies heavy use of Siberian oil and gas and extensive dependence on nuclear power in the European USSR, though a large expansion of output from the Kansk-Achinsk and Ekibastuz coal basins also emerges from the process. Attainment of the optimal plan, however, is considered problematical, since the costs of oil and gas from Siberia may turn out to be higher than forecast, or it may not be possible to develop these resources as rapidly as the optimal plan implies. The considerable share derived for nuclear power also involves uncertainties about how fast the necessary equipment can be produced and what secondary bottlenecks may arise in the industrial branches supplying the inputs for nuclear power stations. (The sectoral impact of investment in nuclear plants is said to be appreciably different from that for conventional stations.) The possibility of developing the implied coal output seems less subject to doubt because the extent and location of reserves are known, and extrapolation of the economic indices is subject to less uncertainty. To allow for possible disappointments in oil, gas, and nuclear output, the modelers make a second iteration with some output constraints for those risky resources and reoptimize with some deterioration in the value of the objective function. This more cautious version the authors call the "effective plan."

Another consideration is that the capital requirements elsewhere in the economy induced via interindustry flows vary from scenario to

*An example of this kind of product is ANSSSR, 1973. It gives shadow prices recommended by the Gosplan for planning decisions about fuel use.

scenario. Specifically, the lower cost plans imply very large inputs of line pipe and other steel goods to meet the oil and gas output targets and hence a need for new capacity in the metallurgical industries. The cheaper plans also imply the need for large additional capacities in the industries producing equipment for nuclear stations.* In any case, another step down from the "effective plan" is then made to ease the burden in some of these bottleneck sectors, and the result is described in the article as the "rational plan."

The features common to all the plans (such as the advantageousness of Siberian oil and gas), together with their differences (such as the cost savings in the objective function that will flow from rapid nuclear buildup), then suggest some general strategic objectives for fuel policy. Concretely, this particular exercise generated three strategic tasks, which the authors describe as follows:

> (1) develop as first priority the oil and gas resources of Siberia (raising output above the current all-Union level) and create the necessary system of pipelines, (2) guarantee the commissioning of nuclear power station capacity adequate to cover the increment of electric power consumption in the western and central regions of the European USSR up to and including the Volga region (except for non-baseload capacity, which will be covered by specialized peaking and maneuverable electric power stations) ; (3) create a Kansk-Achinsk fuel and energy node with an output exceeding that of the Donbass and construct large capacity thermal power stations and enterprises for the technological processing of tens of millions of tons of coal to obtain smokeless semi-coke, synthetic gas and liquid fuel. [*Planovoe khoziaistvo*, 1975:2, p. 35]

Obviously this plan has several other strategic implications not fully underlined by the authors. Most important, there will be a massive flow of energy from East to West, and the planners must concern themselves with means for handling this.

The role to be played by energy exports over the long run is not explicitly addressed in this forecast and, indeed, is usually treated gingerly in this literature. But it is clear from other analyses that the Soviet planners expect to avoid having to import energy and feel it important to maintain a considerable flow of exports, both to meet the energy needs of the fuel-poor countries of Eastern Europe and to earn hard currency. Such an exercise obviously leaves room for debate

---

*As an interesting aside, the authors say that their model did not consider the possibility of importing this equipment. This is a rather strange position to take, since Soviet plans envisage large imports of pipe and pipeline equipment and nuclear power station equipment precisely to help with those bottlenecks.

and requires much more refined analysis before it can be translated into actual energy plans, but it does seem to have been the actual basic rationale for current energy priorities.

This overall strategic vision ultimately becomes the basis for energy R and D planning. Because it grows out of a maximizing model, it is easily manipulable to suggest the areas where the solution of technical problems will have significant payoffs. For example, the massive transport expenditures on gas associated with the priority assigned to west Siberian gas direct attention to the desirability of technological improvements that could ease the resource drain current technology will involve. The emphasis on nuclear power in the western regions suggests all kinds of technical problems, large and small, to be solved, such as the adaptation of nuclear energy to purposes other than power generation and development of the peak load technologies needed to supplement this essentially baseload power source.

How the translation of energy policy goals into an R and D program is made will be further explained in the next chapter.

INPUTS TO THE ENERGY SECTOR: PRODUCTIVITY CHANGES

To conclude this survey of the place of the energy sector in the Soviet economy a few comments on the input requirements of the sector are in order. The energy sector places heavy demands on the resources available to the Soviet planners. Data on employment, fixed capital, and investment available in the standard Soviet statistical handbooks reveal a number of major features.

These are capital-intensive branches, accounting in recent years for almost 30 percent of all fixed assets in the industrial sector. For earlier years this ratio was much smaller—about 20 percent in the interwar period. The rising share may be interpreted as a consequence of several factors. Most important, fuel output has grown faster than national product, and various structural changes have had adverse effects on capital intensity—the rising share of transmission to generation, the resort to lower quality coal resources, etc. We would expect the many other changes that have taken place to have resulted in significant capital savings. The shift to oil and gas, economies of scale in power generation, and the rising share of open-pit mining in coal are examples. In the light of these capital-saving effects, the others must be all the stronger. One explanation of the huge capital costs must be the commitment to hydroelectric power. At present the fixed assets of hydroelectric stations constitute almost three percent of all industrial fixed assets or 10 percent of all fixed assets in the energy sector. This helps explain why Soviet planners were finally forced to reevaluate the high priority they had assigned to hydroelectric power.

The share of new investment absorbed by the energy sector is even higher than its share in fixed assets. The drain on investment resources was particularly heavy in the years after the Second World War until the shift was made to oil and gas. The energy sector took 40 percent of all industrial investment in 1950, for example.

With respect to labor as well, the energy industries are voracious claimants on resources. Total employment in the energy sector at the beginning of the seventies was over two million people, and energy's share in the industrial labor force has generally been around 7–9 percent. There was a tendency for the share to creep up just before the big switch to oil and gas, another indication of the crucial importance of that switch in Soviet energy history.

A very large share of energy employment is in the coal industry, which reported over a million production workers in the fifties and sixties. The United States produced about the same amount of coal with 150 thousand employees, and the U.S. energy sector as a whole has taken a much smaller share of the total industrial labor force than the Soviet energy sector. There are some difficulties in getting figures for those employed in energy in the United States (especially in electric utilities), but in the sixties, when the United States was nearly self-sufficient in energy, about five percent of the industrial labor force was required to provide for its energy needs. And, when allowance is made for the fact that in the United States much of that energy output was processed to a much higher quality level, the difference is still more striking.

As we earlier rejected the notion that the Soviet Union is a country with cheap energy, we also reject the notion that the rapidity of Soviet growth has been aided by the easy expansibility of energy supply. Meeting the needs of its growing economy for energy has placed a heavy drain on Soviet resources; together with the tendency for energy requirements to grow faster than all output, this underlines the crucial importance of the question of productivity and technical change in the energy sector.

2   Soviet Energy R and D

As indicated in Chapter 1, the elaboration of an overall strategic concept of energy priorities sets the stage for working out an R and D program to solve the technological tasks implicit in such a concept. This chapter undertakes to describe how this vision is translated into R and D programs to be carried out by the R and D establishment in the USSR. It also describes the current status of the R and D establishment in the energy sector and makes some summary comparisons concerning the size and composition of the Soviet energy R and D program with that of the United States.

THE PLANNING OF ENERGY R AND D

In addition to its basis in energy policy, the Soviet energy R and D program also flows out of the general process of planning R and D in the Soviet economy.* The Soviet approach to R and D planning involves a mixture of direction from above and initiative from below. The top-down function is in the hands of the State Committee for Science and Technology, the Academy of Sciences, and Gosplan, which are responsible for establishing a set of mission-oriented research programs. In the words of one source: "In the section of the National Economic Plan *'Planning the Basic Scientific-Technical Problems'*, already at the stage of establishing the basic direction of the development of science and technology for the Five Year Plan, the GKNT [State Committee for Science and Technology] together with the Gosplan USSR and ANSSSR [Academy of Sciences of the USSR] sets up a list of basic scientific-technical problems. At the same time the ministries and departments responsible for solving each problem are designated. That list as a rule includes the most important problems of an interbranch character" (Oznobin, 1975, p. 197). For each of these problems, the most important associated tasks (*zadaniia*) are established, and it is these tasks that are actually specified in the Five Year Plan document in the form of particular machines and systems of machines to be

*Several excellent treatments of the Soviet R and D system as a whole are: Zaleski, 1969; Nolting, 1973, 1976a, 1976b; Nimitz, 1974.

created and mastered, technological processes to be created and mastered, and improvement of methods of planning, organization and administration of production (Gosplan SSSR, 1974, p. 11). The construction of a list of *problems* seems to be part of the Five Year Plan process, while the translation into "tasks" takes place both in the Five Year Plans and in the annual plans. This system of designating the problems apparently began with the 8th Five Year Plan, 1966–1970 (*Ekonomicheskaia Gazeta*, 1976:35, p. 8).

The problems are supposed to be grounded in forecasts of technological potentials and economic needs, which GKNT is also charged with organizing. What such forecasts look like can be judged from an exceptionally informative book, *Osnovnye napravleniia nauchno-tekni-cheskogo progressa* (Moscow, 1971), edited by A.S. Tolkachev and I.M. Denisenko and produced by the Scientific Research Institute for Economics under Gosplan. For energy, this book contains a 14-page chapter outlining a number of energy policy objectives and discussing particular technological means for attaining them—such as the use of small diameter and diamond bits to raise drilling speeds and the creation of larger boiler-turbine-generator sets to cut labor expenditure rates.

Some of the problems may be capable of solution within an individual ministry, but most of them are "complex" in the sense that they involve cooperation among a large number of research and production institutions in several ministries. Complex problems are taken care of by "coordination plans," for the working out of which the State Committee is responsible. In an analysis of the 246 complex problems that were established for the 8th Five Year Plan, we find such energy R and D examples as "creation of the production engineering and production of equipment for liquifying natural gas" and "creation of equipment for high speed drilling of deep and superdeep oil and gas wells" (Kosov and Popov, 1972). The "coordination plans" are apparently not part of the official state plan; rather their function is to see that the various tasks (*zadaniia*) that *are* part of the plan form a coordinated solution in each problem area.

This system was changed somewhat for the 10th Five Year Plan (1976–1980) with the substitution of "programs" for the coordination plans. This is supposed to be more than a mere shift in terminology in the sense that the "programs" are to be more explicitly focussed on end results such as the creation of a prototype, construction of a pilot plant, series production, or commissioning of a commercial production facility embodying the new technology (*Ekonomicheskaia Gazeta*, 1976: 35, p. 8).

The list of problems or programs includes basic research as well as applied problems. The "problem" view is characteristically mission-oriented but envisages as a distinct mission the creation of a backlog (*zadel*) of new knowledge that can be used in a future period as the basis for new applied missions or new solutions to existing problems.

The set of mission-oriented tasks we have been describing covers only part of all Soviet R and D—the coordination plans accounted for only 40 percent of total R and D expenditure in the 8th Five Year Plan, for example. The remainder of the national R and D effort consists largely of projects that originate at lower levels, some of which may be very substantial and important, but which are directed toward branch or local problems and require less elaborate interdepartmental cooperation.

This Soviet approach to planning energy R and D shows some similarities with ERDA's planning of U.S. energy R and D (as described in ERDA, 1976). In both cases there is a split between expensive, long-range, uncertain projects larger than can be handled by the resources of individual organizations and smaller, more routine, less risky kinds of work. In the U.S. case, this corresponds in general to a public initiative/private initiative split. The ERDA approach envisages that much of the energy R and D needed to solve energy problems will be conducted by private firms from their own funds and that its own responsibility is to see that long-range, risky efforts that are socially justified but would not pass the test of private profit are not overlooked. In the USSR it is also expected that much research and innovation can be left to the departmental level and that high-level attention should be reserved primarily for novel technologies with long lead times, high-risk effects, or input requirements that go well beyond what individual organizations can handle. I have found no clues as to what share of the total energy R and D program is covered by the "coordination plans," but it would not be surprising to find that, as in R and D as a whole, half or more is determined at lower levels.

As with energy policy generally, the Soviet planning hierarchy seems to lack a node that looks at the energy R and D program as a whole and as its major responsibility in the way ERDA does. Judging from the institutional provenance of the Soviet forecasts we see, much of the initiative comes from the ministries.* There is no energy section as such in GKNT. There is a department of electric power and electric equipment (*otdel energetiki i elektrotekhniki*) and departments for machine building and transport, but none for energy in general. What GKNT does have apparently are ad hoc councils for each interbranch development problem. The interbranch problems in basic research have comparable councils under ANSSSR. These councils pull together specialists in the given area from many different institutions to act as an expert evaluative and policy body for that problem. Most of the literature suggests that the R and D plan for energy is worked out by a process of negotiation among all these groups, presumably with the Gosplan and GKNT having the last word.

MODELING AT LOWER LEVELS

One way to summarize what has already been said is that the R and D desiderata emerging from overall energy policy and the R and D planning process come together at the next level below the central planning bodies in the hierarchical structure in a set of R and D tasks formulated largely in terms of ministerial or broad technology-area clusters. For Minugol', the implication of the "develop eastern coal" strategy is that it must concern itself with technological improvements in stripmining that coal. That goal also implies for Minenergo that it must deal with a complex of problems associated with utilizing that coal in mine-mouth plants and transmitting the power to markets.

The responsibility at this level, however, is not only to translate the rather grossly expressed technological objectives chosen at the center into more fully specified R and D programs and projects, but also to repeat the whole process of establishing a strategic framework of policy goals and shaping R and D inputs to achieve these goals. The ministry has a large agenda of technological tasks growing out of its own operational concerns. The job of the ministry R and D planners is thus to organize and allocate the R and D resources under the ministry's control into an integrated program that supports both agendas.

As this disaggregation proceeds still further and as the R and D process unfolds over time, there is obviously an ever increasing problem of balancing the need to keep R and D efforts focused on the attainment of ministerial goals and the need to change these priorities

---

*The results of one of the studies produced by the power industry are presented in *Energetik* (1970:7, p. 36).

and evaluations as the higher level considerations from which they stem are increasingly modified. Decisions as to the relative priority to give Siberian coal versus Siberian gas in meeting European energy needs must be kept consistent with the technological prospects and research priorities on long distance power transmission versus improvements in pipeline technology. And if the interfuel competition turns out to favor sequential development of the alternatives, the basic versus applied balance within each area must be correspondingly adjusted. For example, judgment that development of long distance power transmission technology will not reach fruition in time to permit reliance on Siberian coal in meeting near-term needs should both induce a heavy quick-fix effort for improving pipeline technology for the enhanced near-term role for gas and a corresponding bias in transmission R and D toward seeking fundamental breakthroughs by means of basic research (on cryogenic approaches, say) rather than a brute force attack on refining traditional approaches.

Strategic R and D goals become widely spread throughout the system as they are translated into R and D programs. Minugol' is instructed to create the technology for efficient open-pit mining of Kansk-Achinsk coal; Minenergo is assigned the task of creating new boilers to burn it and coal processing equipment to upgrade it to a transportable fuel. Others are charged with developing a way to haul it. Within each of these hierarchies, these tasks are further broken down and assigned to R and D organizations. But as this disaggregation takes place and as the R and D work moves forward at the level of the performers, someone needs to stand above the process, repeating that the various subtechnologies—for burning the Kansk-Achinsk coal in boilers, the energy-technological processing technology, and slurry pipelines—are not independent, but have strong complementary or competing relationships and should be designed, evaluated, and developed in that light as well as in their role as means toward ministerial objectives. This is partly a ministerial job, but its interbranch aspects fall on GKNT. Without trying to illustrate with energy R and D specifically, we can say that the State Committee's capacity to carry out this coordination function is seen by Soviet commentators as greatly overburdened by the complexity of the task (Sominskii and Bliakhman, 1972, pp. 8–10).

As the process moves down the hierarchy, the modeling of energy choices becomes more concerned with making operational decisions rather than forecasts and plans, more constricted with respect to technological alternatives, more short-term than long-term, more localized. R and D planning turns into R and D management and becomes ever more intertwined with decisions about current production, investment, operations. But even if those decisions cease to look so much like R and D decisions, they still have an important influence on the sector's performance in innovation and R and D effectiveness. R and D effectiveness is determined not only by what scientific breakthroughs are made and how good the design decisions based on them are, but also on what the producers, the investment decision makers, and the operators do about producing and using the new equipment and ideas.

One of the most important stages in the whole decision-making process is what the Soviet planners call project making, a process in which investment planners make decisions about technologies to be used in some project. At this level there is a great variety of models for decision making, many of them rather standard and mostly what would be called in the United States cost-effectiveness calculations. For example, investment projects are usually evaluated by standard project-making procedures using the pay-out period calculation. As an illustrative case, some higher level organization may define a need to move a proposed volume of oil from Tiumen' to the center, and a project-making organization chooses such variables as pipe diameter, pumping equipment types, and capacities by trading off reductions in operating cost versus the associated capital increments. These design calculations often go far beyond simple one-facility cases and are handled on a branch basis. For example, the gas industry planners try to optimize not only individual pipeline designs, but the design of a whole network, over time, including such aspects of its design as the sequence of constructing alternative routes and the timing of introduction of compressor capacity to arrive at design capacity, by use of dynamic programming models. There are similar models for calculating the effectiveness of proposed new technologies. Much of what the energy planners do follows standard Soviet practice and shares the peculiarities of all such decision making throughout the Soviet economy. The greatest weaknesses in this area probably lie in the limited number of variants usually scanned, and the misleading nature of prices. One of the currently popular themes in Soviet discussions of this modeling is the desirability of using the "systems approach," which tries to avoid the errors of suboptimization by absorbing the interactions of a number of separate issues into a single modeling process. An illustration would be a model that sought to optimize the

development of the nuclear power sector as a whole—including competing technologies, the external fuel cycle, etc.—all in one model,
rather than a model that simply chose between two thermal-neutron
technologies on the basis of cost. But I suspect that despite much lip
service being paid to this ideal, an integrated systems approach is not
widely used in practice.

Finally, as we shall see in examining the progress of specific technological areas, one of the most powerful influences on R and D outcomes is the decision-making behavior of the producers who are supposed to embody new technology in equipment. It often happens that
even when technological improvements have been translated into new
designs, brought to the working stage in the form of prototypes, and
tested in industrial application, their introduction is frustrated by the
peculiar incentive system that inclines the producers away from nov-
elty and from production assignments that are sidelines to their principal specialization.

In short, though one may think of R and D primarily as being influenced by the budgets, goals, and behavior of R and D organizations,
he soon finds that R and D, its translation into new and improved
technologies, and its introduction into the economy is less a product
of the R and D institutions themselves than of the overall setting
within which they work and of a complex set of decisions only part of
which ostensibly concern R and D. Still, the organs primarily charged
with initiating the creation of new technologies are the R and D establishment, and it is to a description of this establishment that we
now turn.

THE RESEARCH AND DEVELOPMENT ESTABLISHMENT

Careful search of Soviet literature reveals a great deal of information on the R and D organizations in the energy field. I have compiled
what I believe is a quite complete inventory of these organizations,
including a great deal of information about their activities, employment, and expenditures. That inventory is too cumbersome to be presented here, but on the basis of that inventory the remainder of this
chapter attempts to—

summarize some general features of Soviet R and D important in
understanding energy R and D;

provide a summary description of the major elements in the energy R and D network and present estimates of their expenditures;

compare the results with U.S. expenditures on energy R and D.

Institutional Description and Concepts.

In the USSR, R and D has traditionally been performed in specialized institutes located at a fairly high level in the administrative hierarchy and somewhat divorced from the enterprises that produce new equipment or utilize new technology. In general, these institutes fit into three main networks.

(1) The most distinctive units of the system, the institutes under the various Academies of Science, represent the highest level of prestige in the system and are supposed to specialize in basic theoretical work. The Academy system includes, in addition to the Academy of Sciences of the USSR, Academies of Science in each of the 14 Union Republics other than the RSFSR. Before 1963, the Academy system also did a great deal of production-oriented research and development, but a reform in that year moved many of the applied research institutes, including a number important in the energy field, out of the Academies into the branch ministries.

(2) Branch institutes tend to have a much more mission-oriented approach than Academy institutes, with responsibility for doing the research and development work needed to meet the technological challenges involved in the development of the branch. But some of these branch institutes are very large organizations, and can devote much of their effort to quite basic research at the frontiers of science and technology. Examples are some of the nuclear power institutes under the State Committee for the Utilization of Atomic Energy, or the Electric Power Institute (ENIN) of Minenergo.

(3) Educational institutions play a much less significant role in the R and D system of the Soviet Union than in the United States, though in recent years it has been an important objective of science policy to strengthen the R and D efforts of educational institutions. Under this policy educational institutions have been given a dual role—they are expected both to improve their theoretical work and to work closely with industrial sponsors in performing industrial problem-solving research through contracts or in industry-sponsored labs. Educational establishments do make a significant contribution to energy R and D.

One encounters a great many difficulties of interpretation in utilizing the available information on Soviet energy R and D expenditures and employment, and it seems only honest to describe how these problems have been handled; others may then be better able to judge the comparability of the Soviet figures presented with U.S. concepts in meaning and coverage. The ambiguities encountered also raise questions that need further study in our effort to understand the meaning and coverage of Soviet aggregate data on R and D.*

First, a significant volume of what seems clearly to be R and D takes place in organizations whose relationship to standard Soviet definitions of R and D expenditures and employment is unclear. Energy R and D involves significant pilot or demonstration projects (*opytno-promyshlennye ustanovki*) such as the U-25 MHD facility, a prototype combined-cycle unit installed in an operating power station, or the pilot installations for energy-technological processing of coal. The operation of these would seem to be R and D and their original cost would be investment in R and D. As one applies the standard Soviet definitions, however, it is not clear that their personnel would be treated as employed in "science," their outlays as expenditures on science, or their capital cost as R and D investment. These could be significant omissions—it is said that the U-25 MHD test plant cost over 100 million rubles to build (*Electronics World*, 15 August 1973, p. 25). In studying energy technology I have been impressed with the importance of the *pusko-naladochnye* (start-up and trouble-shooting) offices charged with getting a new installation (such as a nuclear reactor) to function. The evidence suggests that their work is not considered R and D, but since what they are usually doing is remedying technological faults in incompletely developed equipment, perhaps it should be.

In the standard Soviet descriptions of R and D concepts, institutions engaged in *proektirovanie* (project making) are not generally considered to be engaged in R and D. Many of these project-making institutions, however, include scientific research divisions and in general do a lot of work that would seem at least superficially to be R and D. *Teploenergoproekt* (one of the project-making organs in Minenergo), for example, has its own scientific research institute (NII) and also has

---

*The relative size of Soviet and American R and D expenditures is becoming an increasingly important concern in the light of the way Soviet R and D expenditures are outstripping U.S. expenditures, but there remain many unresolved issues in assessing how comparable the two sets of statistics are. For a statement of concern about the lagging U.S. position see National Science Board, 1975. Several of the general works cited earlier describe the controversies concerning the comparability of the two sets of statistics.

responsibility for the design of new nuclear power plants. In Soviet conditions, the latter involves much more than routine design of a station for which equipment is available and all the technological problems are already settled.

Many organizations have names that indicate responsibilities for project making along with responsibilities for R and D. The network of regional organizations in the oil and gas industries, for example, go under the name of NIPI, (*nauchno-issledovatel'skii i proektnyi institut* —research and project-making institute). I believe these organizations do have responsibility for the main projectmaking function in oil and gas production—i.e., working out the technological scheme and actual engineering decisions for producing an oil field. (Whether they produce the working documents for construction of the above-ground facilities at the field is not clear.)

The word *proektnyi* is used not only to cover the architectural and engineering work on construction projects, but also to mean the design of equipment or of processes. In this sense, it does not differ greatly from what is meant by *konstruktorskaia rabota* (design work) as in the design of a new boiler, an offshore drilling rig, or the circulation pump for a nuclear power plant. In general, the Russians consider this kind of work development, and I treat as R and D organizations many that have the *proektnyi* element in their name—NIPI, NIPTI (*nauchno-issledovatel'skie i proektno-tekhnologicheskie instituty*), and PKO (*proektno-konstruktorskie organizatsii*).

My working hypothesis is that, despite the amalgamation of several functions in an institution, the system often manages to maintain the distinction between R and D and other kinds of activity by the internal structuring of the organization or by separate reporting. Thus I assume that, when it is reported that there are a certain number of persons engaged in "science" in some ministry, these would include the personnel of distinct R and D organizations within big project-making organizations. Similarly, when some organization is said to have spent some given amount for NIR and OKR,* I assume that some attempt has been made to make these numbers conform to the standard concepts of the Central Statistical Administration.

Work on development of new technology performed in production

---

*These standard Soviet acronyms stand for *nauchno-issledovatel'skie raboty* (scientific-research work) and *opytno-konstruktorskie raboty* (experimental-design work), which together are usually taken as meaning the same as the U.S. concept of research and development.

enterprises is not, in general, captured in Soviet statistics or in my inventory. There is a large number of these organizations in the Soviet economy, as indicated by a summary in the Ukrainian handbook for 1974 (TSSU UkrSSR, *Narodnoe khoziaistvo Ukrainskoi SSR*, 1974, p. 90). It shows, for example, in addition to labs and experimental bases, 151 *konstruktorskie organizatsii* that were integral or semi-independent units in industrial enterprises of the electric power and fuel sector in the Ukraine alone. There were another 2,502 KO in machine building enterprises in the Ukraine, a considerable number of which produce such energy equipment as coal-mining machinery. The average size and expenditures of these units are unknown but are surely large enough to imply a considerable volume of energy R and D performed outside what is defined in the USSR statistics as the R and D sector. It is usually said that enterprises do little R and D work, but both Western and Soviet commentators have objected that in fact a great deal of effort must be put in by enterprises to "master" the innovations that have supposedly been prepared by the R and D institutes. A machine may have been designed all the way to production of working drawings by a specialized ministerial KB, but the enterprise that is to produce it may have to redesign the item to make it producible. Most enterprises have an *otdel glavnogo konstruktora* (Department of the Chief Designer), often very large, and it is my impression that in many cases this department does a lot of R and D work. This seems to be the case for walking-dragline excavators, for example, for which most of the design, experimental work, and testing is done in the producing plants. It is often ambiguous as to whether or not the unit at the plant doing such work is kept administratively separate and treated as an R and D organization for purposes of compiling R and D statistics.

In general, my conclusion from looking at energy R and D in detail is that there is probably very inconsistent coverage of the various activities one might want to consider R and D, especially at the development end. The work of administratively separate KB is conceived by the Russians to be R and D, but a great deal of similar work is done elsewhere and does not get into official totals for R and D or into my inventory. If the concern is comparability with U.S. concepts, however, rather than internal consistency, I would judge that these omitted outlays and much of what is done in KB or in NII (and so reported as R and D) is routine engineering design that is excluded from R and D under the definitions enunciated by the U.S. National Science Foundation. In practice, of course, a great deal of similar work may get into U.S. R and D totals.

A final problem has to do with the kind of personnel included in R and D employment. Since expenditures on R and D are heavily concentrated on salaries, the scope of R and D expenditures depends largely on who is considered to be engaged in or supporting R and D activity. In Soviet statements about R and D, employment is described in several concepts. One is total employment in R and D institutions, which includes a considerable number of people engaged in support functions little related to actual R and D—administration, transport, maintenance, information services, and many others. Since R and D expenditures are basically conceived of as expenditures of R and D organizations, all these salaries are included. In Soviet R and D there are said to be only 20 "scientific workers" for every 80 employees in other (i.e., support) activities, while the corresponding ratio in the United States (as shown in the data reported by the National Science Foundation, NSF, for R and D performed in industry) is about 50/50. Soviet expenditure totals seem on this account to be inflated compared to U.S. figures.

A second employment concept is "scientific workers" in R and D organizations, which is supposed to include only those actually conducting scientific research. The goal of this concept is analogous to what the NSF is trying to capture in "scientists and engineers engaged in R and D," though I believe the Soviet concept may be somewhat more restrictive. But it should be remembered that R and D expenditures in NSF statistics pay for a much broader range of personnel than are included in the concept of "scientists and engineers engaged in R and D." Thus, the data for scientific workers in various Soviet energy R and D jurisdictions represent too narrow a base for any effort to move from personnel engaged in R and D to a wage total to an expenditures total. In what follows, accordingly, I work with the broad concept of employment, though this may inflate estimates of R and D expenditures.

Estimates of R and D Expenditures and Employment

On the basis of the inventory of R and D establishments mentioned earlier, it is possible to estimate with varying degrees of accuracy Soviet expenditures on energy sector R and D. In some cases, statements of employment or expenditures are given for whole ministries and, in many other cases, for individual research organizations. The ministerial statements also often give the number of various kinds of institutes in the system, which provides a basis for extrapolating from information on known institutes to aggregates for the system as a whole. Another method of filling in gaps or checking other estimates is to use some fairly systematic relationships between different categories of employment in R and D organizations and between employment and expenditures. One of the biggest problems is to bring all this scattered information to a common time period. In my estimates, 1975 is used as the reference year.

The following section describes the main elements in the energy R and D network and explains the derivation of estimates of their employment and expenditures. The estimates for the network as a whole are summarized in Table 2-1. A reader not interested in the details of estimation should skip directly to the next section, "Comparison with U.S. Energy R and D Expenditures."

MINISTRY OF GEOLOGY (MINGEO). A fairly full description of the ministry's R and D institutes as of 1968 (*Ministerstvo Geologii SSSR*, 1968, pp. 95–101) makes it possible to identify those which specialize in oil and gas. Of the total of 36, 16 were concerned almost exclusively with oil and gas; two others dealt with general regional problems and the economics of exploration generally. Much of the work of both must have been related to oil and gas. Five other institutes were responsible for research and development of equipment and techniques for exploration, a very large share of which must have been for the benefit of oil and gas exploration. Counting those seven institutes as about two-thirds devoted to oil and gas, would give oil and gas 21 of the total of 36 institutes. I imagine the oil and gas institutes tend to be larger than those concerned with other minerals, so that about three-fourths of the total effort should be allocated to oil and gas.

Employment in Mingeo's R and D institutes at the beginning of 1968 was 29,653 persons of whom 10,156 were scientific workers (*nauchnye rabotniki*). Within that subtotal, 244 had "doctor of science" degrees and 2,142 had "candidate of science" degrees. The total number of scientific workers had grown from a figure of 4,491 persons on 1 January 1961, or at an average annual rate of 12 percent.

By 1 January 1975, the network had changed somewhat, with various consolidations and reorganizations, so that the total number of institutes was 34 (*Ekonomicheskaia Gazeta*, 1975:39, p. 2). At that date doctor and candidate degree holders employed in Mingeo were 400

TABLE 2-1. Estimated Expenditures on Energy R and D in the USSR, 1975

Ministry or Department		Employment	Scientific Workers	Doctors and Candidates	Wage Bill (million rubles)	Total Exp. (MR)	Ratio of (5) to (4)
		(1)	(2)	(3)	(4)	(5)	(6)
Mingeo Oil and Gas		44,479	—	—	87.5	175	2
Minneft'		25,000	7,000	1,720	49.2	90	1.82
Mingaz		16,000	—	—	31.5	63	2
Minneftegazstroi		2,500	—	—	4.9	9.8	2
Minneftekhim	A	39,000	—	—	76.8	190	2.47
	B	41,000	—	—	80.8	199	2.47
Minkhimmash	A	18,000	—	—	35.4	60.2	1.7
	B	20,000	—	—	39.4	79	2
Academy Oil and Gas							
Chemical		5,000	—	—	10	20	2
Geological	A	4,500	—	—	8.5	17	2
	B	7,500	—	—	14.5	29	2
Other Acad.		800	—	—	2	4	2
VUZ Oil and Gas		5,300	—	—	10.2	20.4	2
Minugol'		28,000	15,000	2,024	55	92	1.67
Coal Machinery		14,000	9,688	—	27.5	55	2

Academy Coal		—	—	—	—	—	22	—
VUZ Coal		—	—	—	—	—	6	—
Minenergo		12,000	—	—	25	—	50	2
Minelektrotekhprom	A	28,000	—	—	55.1	—	89	1.62
	B	35,000	—	—	68.9	—	111	1.62
Minenergomash	A	15,000	—	—	30	—	60	2
	B	20,000	—	—	40	—	80	2
ANSSSR, Electric Power		—	—	—	—	—	15	—
ANRep, Electric Power		—	—	—	—	—	21	—
VUZ Electric Power		—	—	—	—	—	25	—
Nuclear	A	35,000	—	—	—	—	68	204
	B	37,000	—	—	—	—	73	219
Fusion	A	11,000	—	—	—	—	22	66
	B	12,000	—	—	—	—	24	72
Miscellaneous		—	—	—	—	—	10	—
TOTAL	A	292,579	—	—	—	—	1,298.4	2.08
	B	313,579	—	—	—	—	1,395.2	2.08

A represents low estimates, B larger estimates.

and 4,500 respectively, or 2.05 times the number noted above for 1968 —a growth rate of a little over 9 percent per year. The 2.05-fold increase should probably be discounted somewhat because the 1975 figures refer to Mingeo as a whole rather than to R and D institutes alone, and for the fact that the ratio of these highly trained persons to all scientific workers and to total R and D employment may have risen somewhat. Tending to offset this is the likelihood that the emphasis on oil and gas in the system was even greater in 1975 than in 1968. Thus, I estimate oil and gas R and D employment by simply taking 75 percent of double the number employed in all R and D institutes in 1968, which turns out to be 44,479 persons.

I have found no statements as to R and D expenditures in Mingeo, but expenditures can be estimated as double the wage bill—i.e., employment times the average wage in the science sector in 1975 (155 rubles per month—TSSU SSSR, *Narodnoe khoziaistvo SSSR*, 1975, p. 547) adjusted for social insurance taxes (at the rate of 5.8 percent of wages—Syrovarov, 1966, p. 84). The result for R and D expenditures for oil and gas in Mingeo in 1975 is 175 million rubles.

MINISTRY OF THE OIL INDUSTRY (MINNEFT'). According to a statement made in mid-1975, there were 11 specialized NII, 5 KB, and 17 territorial NIPI in the ministry (*Ekonomika neftianoi promyshlennosti*, 1975:7, p. 4). Total employment in these organizations was about 25,000, of which 1,750 were doctor and candidate degree holders. Total expenditures for scientific research (specifically *nauchnye issledovaniia*) were about 90 million rubles per year. These figures do not include the experimental factory controlled by the All-Union Scientific Research Institute for Drilling Equipment (VNIIBT).

There are some checks supporting the plausibility of these figures. Detailed information is available on several of the regional NIPI suggesting an average budget of about 1.5 million rubles per year, for a total of 25 million rubles. A plausible average for the 11 specialized NII is 4 million rubles per year. The All-Union Research Institute for Drilling Equipment (VNIIBT) had a budget of 4.3 million rubles in 1967 and has surely grown a great deal since then. Except for the All-Union Oil and Gas Research Institute (VNII), it is probably the biggest institute in Minneft'. On the basis of employment, the Institute for the Geology and Production of Mineral Fuels (IGiRGI) must have had a budget of over 4 million rubles in 1975. An average of 4 million rubles per NII and KB plus the 25 million rubles for territorial NIPI gives a total of 89 million rubles.

The wage bill for 25,000 employees would be 49.2 million rubles— 55 percent of the total reported expenditure of 90 million rubles—and is in line with the share of wages in all expenditures in R and D generally.

MINISTRY OF THE GAS INDUSTRY (MINGAZ). As of 1975 this ministry had six specialized NII, six complex NIPI and four NPO containing R and D organs of various kinds (*Gazovaia promyshlennost'*, 1976:11, p. 10). Total employment in "science" in the branch is reported as 10.5 thousand persons, but this probably excludes people in KB or doing OKR. The Research Institute of the Gas Industry (VNIIgaz— probably the largest of the NII) had employment of 3,000 in 1973 (*Gazovaia promyshlennost'*, 1973:8). If average employment in all NII were half that, employment in NII alone would be 9,000 in 1973 and more in 1975. The employment figure for VNIIgaz in 1973 implies a wage bill of 5.7 million rubles and a total expenditure of 10–12 million rubles in this institute alone. One of the new NPO (*Soiuzgazavto-matika*) did NIR and OKR in 1974 of 5.86 million rubles. Both are probably larger-than-average institutions, but assuming that each of the 17 significant organizations of 1972 had expenditures of 4–5 million rubles, the total would be 70–85 million rubles. This suggests that the figure for employment in "science" above is too small, and I use a figure of 16,000 for employment which implies expenditures of about 63 million rubles.

MINISTRY FOR THE CONSTRUCTION OF OIL AND GAS INDUSTRY ENTER-PRISES (MINNEFTEGAZSTROI). This is a small ministry, weak in R and D, and there is little information on its R and D activity. It is possible, however, to identify five or six significant institutes, and, if average employment is assumed to be 500 persons, employment would be 2,500 persons and expenditures about 10 million rubles.

MINISTRY OF OIL REFINING AND PETROCHEMICALS (MINNEFTEKHIM). In the early seventies, total employment in research and *proektnye* organizations was, by one account, 48,000 persons (*Neftepererabotka i neftekhimiia*, 1971:3, p. 5), and by another, 50,000 persons (*Khimiia i tekhnologiia topliv i masel*, 1971:5, p. 2). The number of professionals (by the educational criterion of being ITR, doctor or candidate) was about 29,000, of which 1,238 were doctors and candidates (*Khimiia i tekhnologiia topliv i masel*, 1973:9, p. 3). This statement seems to refer to 1971. During 1966–1970 expenditures on science in this ministry grew at an average annual rate of 6.5 percent per year (Gvishiani, 1973, p. 159), and assuming an employment growth rate of 5 percent in 1970–1975 would give 58–61 thousand persons in 1975. But the 48–50 thousand total includes employees in *proektnye* organizations, in *pusko-naladochnye* offices, and in experimental (*opytnye*) factories. No information is available to separate R and D specifically on petrochemical work from the total, except descriptions for many of the organizations of their major functions. This suggests that a downward adjustment of about a third is in order, yielding an employment in research and development in 1975 of 39–41 thousand persons. I simply translate the estimated employment into a wage bill and then into a figure for total expenditures on the basis of a statement that in this ministry in 1968 the wage bill in R and D was 38.2 percent of R and D expenditures (Grishaev, 1970, p. 95). The estimate of expenditures is thus 190–199 million rubles.

MINISTRY OF CHEMICAL AND PETROLEUM MACHINEBUILDING (MINKHIM-MASH). Much of the R and D for oil and gas is performed in the machine building branches that produce its equipment. Some of this equipment is produced in ministries other than Minkhimmash (e.g., drilling rigs in the Ministry of Heavy Machine Building), but that will be discussed later. This ministry has 48 specialized NII, counting branches (*filialy*), and 25 PKO, both of which are considered R and D organs. I have no information on employment in institutes. To get a number to start with, I assume average employment per organization as 500 people, for total employment of 36,500 and a wage bill of 72 million rubles. In this ministry in 1968 expenditures on science were 1.71 times the wage bill, and assuming the same ratio held in 1975, expenditures would be 122.8 million rubles (Grishaev, 1970, p. 95).

A possible check shows that this may be too large. A decree of September 20, 1976, established the "unified-fund" method of financing in this ministry (*Sobranie postanovlenii Pravitel'stva SSSR*, 1976:22, pp. 409–418). The decree stipulated that the Gosbank should provide a credit of 12.5 million rubles for the first quarter of 1977, presumably pending the determination and transfer of profits to the fund at the end of the quarter (deductions into the fund are fixed as a percent of profits). This would suggest an annual volume of 50 million rubles, though the fund will also receive some budget financing. The decree suggests that total funds will be $\frac{4.4}{3.7}$ times the profit deductions, or 60 million rubles.

For now, however, I use the original approach and further assume that half the work is for oil and gas (rather than for the chemical industry proper). Hence, employment is about 18,000 and R and D expenditure is about 60 million rubles.

INSTITUTES IN THE ACADEMIES AND HIGHER EDUCATIONAL INSTITUTIONS (VUZY)—OIL AND GAS. A large number of VUZ and Academy institutes and institutes in higher educational institutions are concerned with oil and gas. The Academy system makes its biggest contribution in the areas of oil chemistry (including such related fields as physical chemistry and catalysis) and geology.

My inventory shows 15 significant institutes in the Academy system concerned with petrochemistry. These are fairly large institutes, and a plausible average expendure is 2 million rubles, for a total of 30 million rubles. Much of this research must concern such downstream activities as petrochemical production, so I reduce it to 20 million rubles and 5,000 people.

The Academy system is also strong in geology and geophysics. Some 13 institutes that are heavily specialized in oil, gas, and coal can be identified in the Republican Academies and in branches of the ANSSSR. The seven geological and geophysical institutes in the ANSSSR seem less oriented to specific fuel branches, though most of them work indirectly on energy sector problems. I take a figure of 15 institutes as appropriate here. With average employment of 300–500 people, this would mean 4,500–7,500 people and 17–19 million rubles.

There are also a couple of more specialized institutes—the Gas Institute in the Ukrainian Academy, and the Institute for Problems of Deep Fields in the Azerbaidzhan Academy—to add at an estimated expenditure of 2 million rubles each.

There are seven VUZy oriented mostly to oil and gas, and they do a significant amount of research work. The Gubkin Institute (MINKHiGP) is the principal VUZ in the oil and gas field and did over 4 million rubles worth of research in 1967. A figure of 5 million rubles for 1975 is perfectly plausible.

One source says that during the 9th Five Year Plan the seven VUZy in oil and gas did about 90 MR of NIR, or an average of 18 million rubles per year. Given growth and the figure for the Gubkin Institute, 20 million rubles of R and D in 1975 seems reasonable. Since the Gubkin Institute had 51 professors and 350 assistant professors (*dotsenty*) doing research plus 1,100 workers in the research sector, counting the teachers as halftime researchers would give 1,300 researchers for the 5 million rubles of research. I extrapolate this employment/expenditures ratio to the whole collection, giving 5,300 persons employed in VUZ R and D.

MINISTRY OF THE COAL INDUSTRY (MINUGOL'). The most complete statement available refers to 1970, when there were 24 NII, employing 23,000 persons (excluding 4,711 workers in experimental factories) and with a total expenditure on research and experimental work of 53 million rubles. This expenditure figure seems too small, since at the 1970 average wage 23,000 workers imply a wage bill of about 40 million rubles, almost as large as total expenditures, which seems implausible. The Donets Coal Institute (DonUGI), one of the largest institutes, had an annual budget in 1966–1970 of about 3.5 million rubles and employment of about 1,200 persons (implying a wage bill of slightly over 2 million rubles), so that its total expenditures were 1.67 times the wage bill. I suspect we must go by the employment figure in estimating a total for the branch, and if the DonUGI expenditure per worker is representative, 23,000 employees would imply about 60 million rubles. But all this is for 1970, and I adjust employment to a 1975 base by raising it at the rate of about 4 percent per year, which would give over 92 million rubles per year in 1975.

Minugol' also inherited in 1973 the R and D network for coal mining machinery formerly under the control of the Ministry of Heavy Machine Building. In 1967 this network included 19 KB and technological institutes designing coal mining machinery plus 10 experimental factories. The number probably did not increase much by 1975, and an estimated 750 persons per KB in 1975 would mean 14,000 employees and about 55 million rubles of expenditures.

In the Academy network, I count 11 institutes in exploration, processing, explosives, mining, etc., which are almost exclusively or heavily oriented toward coal, and which I will assume to have average expenditures of 2 million rubles for a total of 22 million rubles.

There are six VUZy heavily concentrated on coal—the big ones did 1.5–2 million rubles per year of research work in the sixties. The

Leningrad Mining Institute did 3.8 million rubles in 1972 (only partially devoted to coal). An average of 1 million rubles per VUZ for coal in 1975 would mean 6 million rubles.

MINISTRY OF ELECTRIC POWER AND ELECTRIFICATION (MINENERGO). The ministry had 20 *golovnye* NII, employing 12,000 people in 1975. (*Golovnye* institutes are large organizations, each responsible for a particular area of research, both in the sense of performing it themselves and supervising related work in smaller institutions of lower rank.) The implied R and D expenditure of 50 million rubles seems small to me. The numbers also imply 600 employees per NII, which seems too small for the old and prestigious institutes of this branch.

The minister, P. S. Neporozhnyi, says in a statement that probably refers to 1970, that under the Minenergo organization that runs the whole R and D and project-making show, there are 70,000 employees. Of the five main project-making organizations, Gidroproekt has 17,000 employees and Teploproekt, 12,000. These are the two largest, and the implied 58 thousand persons in project making (the total employment of 70,000 less the 12,000 in NII) would mean average employment of 7,000 in the three smaller *gipros*, which seems somewhat large. Also, the *gipros* themselves contain NII and KB that may not be included in the 20 *golovnye* NII. But, lacking a basis for adjustment, I will use the 50 million ruble figure.

MINISTRY OF THE ELECTROTECHNICAL INDUSTRY (MINELEKTROTEKH-PROM). Most of the R and D work on generators, transformers, switch gear, and other transmission equipment is performed in this ministry. (Turbines, boilers, and nuclear reactors are in the Ministry of Power Machine Building.) It has an elaborate network of R and D organizations, described in 1974 as having 33 specialized NII and KB, of which 17 were subordinated to enterprises and the rest to the ministry. Within the 16 subordinated to the ministry, 10 were KB and the rest, NII (*Planovoe khoziaistvo*, 1974:11, pp. 52–53). There are also numerous smaller R and D organizations—the 1974 source said that there were a total of 70 NII and KB in the ministry with their own experimental facilities, implying that there was still another group without such facilities.

I have been able to find virtually nothing on the size of these institutes or their expenditures. Total expenditure from the unified fund in this ministry in 1972 was about 300 million rubles (Bazarova, 1974), and was supposed to grow at about 9 percent per year in the 9th Five Year Plan (Gvishiani, 1973). But this fund finances not only R and D but also "mastering" (*osvoenie*, putting a new item into production), and we also know that it supplies only about half the funds for R and D work, the other half coming from contracts with customers (Arkhangel'skii, 1976, p. 103). All this implies a very large volume of funds in 1975, something over 700 million rubles (including *osvoenie*). Assuming 70 NII and KB, this would be about 11 million rubles expenditures per institute, which seems impossibly large. For example, it would imply over 3,000 employees per organization.

As an alternative approach, I assume that half the R and D work in this sector is for electric power generating and transmitting equipment (as opposed to energy-using equipment) and that this means 35 R and D organizations. I further assume average employment of 1,000 persons. A study in Leningrad showed planned expenditures per worker in 12 NII and KB in the electrotechnical sector of 2.4 thousand rubles per worker, or 1.62 times the employment cost per worker in the mid-sixties (Bliakhman, 1968, p. 22). Hence, 35,000 workers in 1975 would mean a wage bill of about 68.9 million rubles and 111 million rubles of R and D work. If average employment were only 800 persons, the corresponding figures would be 28,000 workers and 89 million rubles of expenditures.

MINISTRY OF POWER MACHINEBUILDING (MINENERGOMASH). There is little information on the R and D organizations in this ministry, but my inventory shows that in the mid-seventies it had 15–20 NII engaged in R and D on boilers, turbines, and nuclear plant equipment. Some of these are very large institutes (e.g., the Central Boiler and Turbine Institute) and assuming an *average* of 1,000 employees per institute implies 15–20 thousand persons or 60–80 million rubles of expenditure.

ACADEMY INSTITUTES AND INSTITUTES IN HIGHER EDUCATIONAL INSTITUTIONS—ELECTRIC POWER. There is still a large amount of basic research on problems of electric power in the Academies, though most of the applied research institutes were lost in the 1963 reshuffle.

The USSR Academy (ANSSSR) has three divisions concerned with energy R and D: Physico-technical Problems of Electric Power, Physics, and Nuclear Physics. Information as to what institutes actually exist

in the USSR Academy in these divisions is quite inconsistent. At various points in the late sixties, mention is made of seven large institutes in electric power R and D as being in the Academy, but the 1975 Directory of the Academy shows only three: the Institute of High Temperatures in Moscow, the Institute of Thermophysics in Novosibirsk, and the Siberian Power Institute in Irkutsk (*Akademiia Nauk SSSR, Spravochnik*, 1975). The Directory may be incomplete, or more of these institutes may have been lost over the years to the ministries. M. A. Styrikovich, head of the Division of Physico-technical Problems of Electric Power, says that much of the effort of the Division goes into coordinating the work of institutes in the USSR Academy with those in the Republican Academies and those in the ministries. It seems strange that this Division should have under its control only a single research establishment (as shown in the 1975 Directory), but, given the evidence of the Directory, I conclude that there are only the three institutes in the ANSSSR. In view of the fact that some power institutes in the Republican Academies have expenditures of up to 3.5 million rubles, average expenditure for the institutes in ANSSSR should be at least five million rubles.

In the Republican academies, it is possible to identify nine power institutes and four smaller power "departments" (*otdely*) in the smaller academies and regional centers. Employment and expenditures can be estimated for several of these, with expenditures ranging from about one million rubles in the Electric Power Institute (*Institut energetiki*) of the Latvian Academy to 3.5 million rubles each in a couple of institutes in the Belorussian Academy. My estimate for the group as a whole is 21 million rubles; this does not include the institutes in the area of nuclear power.

The biggest of the VUZy, the Moscow Electric Power Institute (*Moskovskii Energeticheskii Institut* or MEI) is a significant center of electric power R and D, and performed 13 million rubles of NIR in 1967. Since then the number of labs has increased from 17 to 23, and, if we adjust expenditures by that yardstick (probably too small as an index of growth), we would have 17.6 million rubles in 1975; I round it up to 20 million rubles in 1975. The other electric power VUZy are much less important contributors to R and D. The Leningrad Polytechnical Institute has a high voltage lab that does almost a million rubles worth of R and D per year; the Leningrad Electrotechnical Institute does a few hundred thousand rubles worth of work per year.

Most polytechnical institutes in the USSR have an electrical engineering department, but on a smaller scale than LPI. I assume a total of 5 million rubles for all VUZy other than MEI.

NUCLEAR POWER RESEARCH. The major R and D organizations in this field are under the State Committee for the Peaceful Utilization of Atomic Energy, but the Academies do a lot of work on fusion, and, to judge from descriptions of the nuclear physics institutes in the Academies, by A. M. Petrosiants (Chairman of the State Committee), and others, a great deal of their work is concerned with nuclear power problems.* Most of the Committee's institutes seem to be very large. The Kurchatov Institute has 5–6 thousand employees, and the Research Institute for Nuclear Reactors (NIIAR) in Dmitrovgrad had 2,000 in 1970. (The Kurchatov payroll and budget is said to have been fairly stable for several years.) The seven big institutes under the Committee probably employ 19–20 thousand persons.

In addition to what the Kurchatov Institute does on fusion research, there are fusion programs in four other big institutes (Physico-technical Institute of the ANUkrSSR in Khar'kov, the Lebedev Physics Institute, the Ioffe Physico-technical Institute of the ANSSSR, and the Physico-technical Institute in Sukhumi). The Khar'kov institute is said to have 5,000 employees and the Lebedev Institute 3,000. We might estimate total employment in the four at 9–10 thousand persons (though some of these people are working on problems other than fusion) and transfer to fusion another 2,000 persons from the Kurchatov total.

The nine nuclear physics institutes in the Academies probably had employment of at least 10,000 persons. Employment in the Institute of Nuclear Energy of the Belorussian Academy was about one thousand in 1975. The Georgian Institute of Physics had expenditures in 1969 of 2,408 MR, which implies employment of 700–800 at that time, so that a figure of a thousand or more is likely for 1975. The average size of institutes in the Academy of Sciences of the Kazakh SSR in 1975 was something over 400 persons. Its nuclear physics institute would probably be larger than the average, but with less than a thousand employees. The other Republican Academy institutes would probably be much smaller, while those in the Ukrainian Academy, the

*A fairly extensive description of nuclear research and nuclear research institutes is given in Petrosiants (1976, pp. 353–403). Numerous Westerners have visited many of these institutes, and their reports contain numerous statements about employment in them.

SOANSSR and the ANSSSR, would be larger than the ones mentioned so far.

The descriptions of the work of these institutes make it clear that they do research on many topics that are only tenuously related to nuclear power; but in order not to leave them out of account entirely, it may be reasonable to show them as contributing 5,000 persons to nuclear power R and D.

The implied total employment figures of 35–37 thousand persons for all nuclear, and within that 11–12 thousand for fusion, work out to very large wage bills, and these have been multiplied by three to get estimates of total expenditures. (I assume that the material intensity of R and D in physics must be much higher than in the other kinds of research discussed so far.)

GEOTHERMAL. The geothermal program is small and scattered over a large number of small units. A Scientific Council on Geothermal Problems in the USSR Academy of Sciences has been in existence since 1964 (Dvorov, 1977, p. 25). A book on geothermal research in the Georgian SSR (Chikradze, 1972) describes the Geophysics Institute of the Georgian Academy as the main research organization and mentions work by the Laboratory of Geothermics and Hydrogeochemistry of the Geological Institute of the USSR Academy. The Leningrad Mining Institute operates a laboratory for work on dry, hot rock. The Institute of Geological Sciences and the Council for the Study of Productive Forces of the Armenian Academy have been doing research since 1957. Minenergo, the USSR Academy, and some other organizations are involved in the power generation aspects; the Academy and Mingeo are concerned with resource evaluation. Mingaz has the major responsibility for operations through four field administrations, and its R and D organizations do research on drilling geothermal wells and on producing and handling the brines. There is no basis for estimating an employment or expenditures figure, but it must be very small.

SOLAR. This program, too, is scattered among many small organizations. The scientific journal for this area of research—*Geliotekhnika*—is published by the Academy of Sciences of the Uzbek SSR, which administers the S. V. Starodubtsev Physico-technical Institute in Tashkent. This institute seems to be a major locus of solar energy research and could be a large institute, since the Uzbek Academy has 3,699 scientific workers in its 31 institutes, and the physics institutes are likely

to be larger than average. Examination of the institutions whose research is reported in *Geliotekhnika* suggests that many institutions
have solar research as a sideline to their main activity, such as institutes of electronics and institutes that need remote power sources (e.g.,
the Tashkent Communications Institute). The Physico-technical Institute of the Academy of Sciences of the Turkmen SSR works on solar
cells, but this must be a very small program, since the average of scientific workers per institute in the Turkmen Academy was only 54 in
1975. There is a Research Institute of Current Sources, under Minpribor, which is a significant performer of this kind of research. There
is still some research on solar energy at the Electric Power Institute
(ENIN), and some efforts at VUZy.

An article reviewing ten years of the journal's activity says that
"there has been a considerable expansion in solar-technology research
in the Soviet Union in the last ten years," and adds that, in addition
to institutions active in 1965 (Leningrad, Moscow, Erevan, Tashkent,
Ashkhabad), new schools have been established in the Ukraine (concerned with the effect of high intensity light and heat fluxes and high
temperatures on matter), in Moldavia (development of semiconductor
photocells and effect of light on seeds), and in Azerbaidzhan (development of autonomous sources of supply using solar energy). They mention a three-meter solar furnace at the Uzbek Academy and a ten-meter
furnace being built at Erevan. According to their summary, in 1975
there were 30 doctors and 100 candidates researching solar energy
(*Geliotekhnika*, 1975:314, pp. 3–4).

OIL SHALE AND PEAT. There is only one institute for the oil shale
industry—the Scientific Research Institute for Shale (NIIslantsev) in
the Ministry of Petroleum Refining and Petrochemicals. *Mining* of
shale is administratively under the coal industry, and some of the coal
industry research organizations have a place in their programs for
shale mining. There is also an Institute of Chemistry in the Academy
of Sciences of the Estonian SSR which works mostly on kerogen, and
had a 1975 budget of about 2.5 million rubles. I have found only two
institutes for the peat industry: the Peat Institute (Institut torfa) of
the Belorussian Academy in Minsk, which had a staff of 232 persons
in 1968; and the All-Union Research Institute of the Peat Industry in
Leningrad, with a Moscow branch. The latter is under the jurisdiction
of the Ministry of Agriculture.

The reader who has followed the description of the evidence and its translation into estimates of employment and expenditure may justifiably express considerable skepticism about the accuracy of the results. Some decisions are admittedly arbitrary; overall, the results are probably biased downward, a choice based on the fact that one use for these totals is to compare them with U.S. expenditures on energy R and D. The totals shown are probably a conservative estimate compared to what would be generated by more complete information about the energy R and D establishment, both because of this bias and because of possible omissions, especially of energy-related work outside the energy sector.

COMPARISON WITH U.S.
ENERGY R&D EXPENDITURES

For purposes of comparison with the United States, it will be useful to summarize the data of Table 2-1 in slightly different form. In Table 2-2 expenditures are grouped in terms of the areas usually considered in discussions of energy R and D expenditures in the United States. Throughout, I have chosen the lower figure when Table 2-1 showed

TABLE 2-2. Soviet Energy R and D Expenditures by Major Programs

	Million Rubles (1)	Percent (2)	Million Rubles (3)	Percent (4)
Nuclear	204	16	204	16
fusion	66	5	66	5
Fossil energy sources	808.4	62	693.2	53
geology	192	15	192	15
coal	163	13	108	8
oil and gas	453.4	35	393.2	30
Solar energy	5	negl.	5	negl.
Geothermal energy	negl.	negl.	negl.	negl.
Conservation	260	20	111	9
electric power	258	20	109	8
other	2	negl.	2	negl.
Environment and safety	16	1	16	1
Equipment development	*	*	264.2	20
Other	5	negl.	5	negl.
TOTAL	1,298.4		1,298.4	

*Distributed by sector.

a range. For the most part these are just obvious sums of items from Table 2-1, but there are a few amendments. I have transferred from the ministerial sums generous estimates for the expenditures of the six institutes explicitly concerned with environmental and safety problems, and for a new institute concerned with conservation, and shown them under corresponding headings. These are surely underestimates of expenditures for these purposes, since some work on these problems takes place in other of the institutes covered and also in institutes outside the ministries examined here. In the first column, expenditures in the equipment ministries are included with the appropriate energy producing ministries, but in column (3) they are pulled out into a separate category. This is an incomplete figure for machinery development, since many of the institutes in the producing ministries are also concerned with designing new equipment. But I see no way to separate out appropriate amounts.

So far as I can find, there exists no satisfactory summary of overall U.S. energy R and D expenditures, but a reasonable total and structure can be put together on the basis of Table 2-3.

TABLE 2-3. U.S. Energy R and D Expenditures, 1975*
(million dollars)

	Federal Performers (NSF)	Industrial Performers (NSF)	Total (1) + (2)	Federal Funds (NSF)
	(1)	(2)	(3)	(4)
Fossil fuel	60.8	524.0	584.3	289.1
Oil	22.0	326.0	348.0	n.a.
Gas	3.1	62.0	65.1	n.a.
Shale	6.6	14.0	20.6	n.a.
Coal	29.2	86.0	115.2	n.a.
Other fossil	0	36.0	36.0	n.a.
Nuclear	412.5	631.0	1,043.5	612.6
Fission	343.3	603.0	946.3	469.1
Fusion	69.2	28.0	97.2	143.5
Solar energy	11.9	11.0	22.9	54.7
Geothermal energy	19.5	5.0	24.5	25.0
SUBTOTAL	504.7	1,171.0	1,675.7	961.4

All other except environment + safety	268.5	95.0	363.5	375.2
Conservation	28.0	n.a.	n.a.	64.4
Basic research	233.0	n.a.	n.a.	291.7
Systems studies	4.1	n.a.	n.a.	0
Other	3.4	n.a.	n.a.	19.1
SUBTOTAL	773.2	1,266.0	2,039.2	1,336.6
Environment and safety	104.1	not covered	n.a.	368.8
Environmental control	5.4	not covered	n.a.	⎫ 265.7
Environmental effects + health	89.7	not covered	n.a.	⎬
Safety	9.0	not covered	n.a.	⎭ 103.1
GRAND TOTAL	877.2	1,266.0	2,143.2	1,705.4
of grand total, in machinery branches		534.0	534.0	n.a.
of grand total, in geology and resource assessment	27.5	n.a.	n.a.	n.a.

*Calendar year for industrial performers, fiscal 1975 for federal data.

SOURCES: Federal performers from NSF, Reviews of Data on Science Resources, *Energy and Energy-Related R and D Activities of Federal Installations and Federally Funded Research and Development Centers*, NSF 76–304, April 1976. Industrial Performers from NSF, *Research and Development in Industry, 1974*, NSF 76–322. Federal funds from NSF, *Analysis of Federal R and D Funding by Function*, NSF 76–325. For federal funds, the allocation of items in the sources by the categories of this table is as follows:

Basic research = ERDA high energy physics and basic energy sciences
Environment and safety

Health and safety research (Mines-DOI)	31.9	(S)
Air quality effects research (EPA)	15.2	
Radiation effects research (EPA)	1.5	
Energy-related environmental research (EPA)	13.9	
Biomedical and environmental research (ERDA)	142.5	
Environmental and fuel cycle research (NRC)	2.5	
Energy-related environmental control programs (EPA)	80.7	
Nuclear materials security and safeguards (ERDA)	6.2	(S)
Operational safety (ERDA)	3.3	(S)
Environmental control technology (ERDA)	8.2	
Environmental effects of energy (RANN-NSF)	1.2	
Reactor safety (ERDA)	0.0	
Reactor safety research (NRC)	60.4	(S)
Safeguards research (NRC)	1.3	(S)
Subtotal safety	103.1	
Subtotal other	265.7	
Total	368.8	

The NSF has published estimates of R and D expenditures by federal performers and by industry for 1975, but in the NSF system this means that energy R and D performed in other sectors (such as non-profit institutions, universities, and so on) is not included. However, NSF also publishes an analysis of federal funds for *financing* R and D, by function, with the functions being more or less assimilable to the categories of the two summaries already mentioned. In this functional breakdown, "environmental" and "basic research" are shown as functions separate from energy, and one has to pull out of those tables the programs that seem energy-related. Since federal funds finance some R and D performed in sectors other than government and industry, in those categories where federal funds exceed the sum of expenditures by federal performers and industrial performers, the former should be taken as a better estimate. The major categories of these are fusion research, basic research, and environmental safety, in which there are significant expenditures in universities (and perhaps in other omitted performing sectors as well).

When the additional 46.3 million dollars for fusion, 58.7 million dollars for basic research, and 264.7 million dollars for environmental and safety research estimated in this way are added to the amounts shown in column (3) the total becomes 2,512.9 million dollars.

This approach will still fail to capture such energy R and D as is performed in sectors other than government and industry and is either self-financed or financed by industry. But I believe those omissions will be quite small in relation to the total.*

Our approach has the advantage of sticking with the NSF concept of current expenditures only (capital expenditures and demonstration plants are excluded) and also permits separating the R and D performed in the machinery branches supplying equipment for the energy sectors. Unfortunately, the method hybridizes fiscal years (for federal funds) and calendar years (for industrial performers), but permits us to use the "obligations" concept for federal funds and performers (rather than the budget authority concept), thus maintaining approximate consistency with the "expenditures" concept used for industrial performers.

Acknowledging that many conceptual differences with respect to

*It is also possible to get another view of federal financing by examining the ERDA or OMB statements of "federal energy R and D" (which is essentially a financing concept); but it turns out this view does not really add any useful information, and I have not included it in Table 2-3.

coverage and internal classification interfere with comparability of the U.S. and Soviet summaries, a comparison of the two still suggests some interesting conclusions and raises some puzzles that need further study.

To assess the overall relative size of the two countries' efforts, some general notion is needed of how much R and D resources a ruble buys. In some other calculations I have made, based on converting the various elements in the Soviet R and D expenditure total at a plausible ruble/dollar ratio, the overall conversion ratio that emerged was 3.8 dollars to the ruble (that is, one ruble buys as much resources as 3.8 dollars). This refers to R and D expenditures in general, as it is not possible to make such a calculation for energy R and D alone. The dominant influence on this result is the relative wages of U.S. and Soviet employees in R and D, figured separately for the two categories of scientific workers and supporting personnel. Labor is the most important element in R and D expenditures, and the dollar cost of labor relative to the ruble cost is so high that the dollar/ruble ratio for all R and D resources is of the order indicated. I stress that any such comparisons as this are extremely treacherous, but even if the ratio were significantly different (say 3 dollars per ruble), it appears that the Soviet energy R and D program is very large compared to the U.S. program. At a conversion ratio of 3.8 dollars per ruble, Soviet expenditures are almost five billion dollars in 1975, about double the U.S. figure for the same year.

In interpreting that statement, however, many qualifications must be kept in mind.

(1) First, there is no doubt a strong index number effect at work here. Soviet R and D input mixes are heavy on manpower compared to U.S. mixes, and since the manpower/other input price ratio is higher in the dollar price system than in the ruble price system, a comparison in the ruble price system would no doubt show a significantly lower relative standing.

(2) The Soviet figure is probably inflated in the sense that it covers some activities that are not considered R and D and are not included in the U.S. totals. As explained earlier, the method of totaling budgets of research *organizations* rather than research *programs* brings in expenditures on employees engaged in all kinds of auxiliary operations that are peripheral to actual R and D and that do not get counted in U.S. totals. The janitors, the chauffeurs, and a top-heavy load of administrators are examples.

(3) Examination of the differing structures of the two totals suggests

that the Soviet total includes some activities that are not counted as R and D in the United States. For example, expenditures on geological R and D are very large in the USSR, accounting for 15 percent of the total, versus the 1 percent we can isolate for the United States. The geology figure for the United States in Table 2-3 covers only federal performers, and it may well be that significant amounts of company financed R and D in oil and gas is for geological and exploration tasks. In the Soviet case, however, our geology figure similarly omits the large amount of geological work done in the R and D organizations of Minneft' and Mingaz.

(4) On the other hand, the Soviet total includes little activity that could be described as safety and environmental R and D, which accounts for almost 15 percent of the U.S. program. Some Soviet environmental and safety work is done outside the network of organizations this survey has covered, and what is done in this network is not fully separated out. Also, in the United States a very large amount of environmental research would seem to be included as energy-relevant only because of institutional peculiarities (ERDA finances a big environmental and health program) or for image purposes. If this category of expenditures is removed from both sides, the remaining 1282 million rubles of Soviet R and D seems an even larger program in relation to the remaining 2144 million dollars worth of U.S. energy R and D.

(5) Some structural differences are so striking that they can probably be taken as reflecting real differences, despite underlying noncomparabilities in definition or errors in estimation. Expenditures in the field of nuclear energy are much more important relative to fossil fuel programs in the United States than in the Soviet Union. The nuclear/fossil ratio is 1.88 in the United States, but 0.25 in the USSR. If we compare the two halves of this ratio separately, the figures suggest that the Soviet nuclear program may be smaller in absolute terms than ours. I do not know enough to judge whether that is plausible. The Soviet program is certainly ambitious and productive in the sense of covering a lot of areas, including fusion, and in the sense of having created an operating technology. It may be somewhat starker in terms of alternatives; perhaps it is more efficient. The small relative size of the Soviet nuclear program is so striking that it makes one wonder if some egregious omission has been made in adding up the elements of the Soviet nuclear program. The institutes I have included are mostly those which Westerners have visited; perhaps there are some important se-

cret ones. It is likely that significant amounts of work are done for the nuclear power program in institutes not covered—in mathematical institutes, chemical and material research institutes, and so on. Whatever work is done by Minenergomash for the nuclear program is shown in my table under electric power rather than nuclear.

On the other hand, I have been fairly generous in estimating outlays for the nuclear program. For example, I have blown up the wage bill by a factor of three rather than the two or less characteristic of other fields, I have included a number of academy nuclear research institutes that are only partially energy-relevant, and so on.

Soviet expenditures on the fossil part of the ratio, however, seem extraordinarily large—five or six times larger than in the United States. In trying to explain these large expenditures, we can point to several contributing factors beyond those already mentioned (such as the inclusion of geology). I believe, for example, that these include expenditures on much activity that is not R and D at all, but routine managerial work, routine production engineering, technical support, and other such functions that may get into the U.S. figures to some extent (though they are not supposed to by the definitions) but that get into the Soviet figures on a much larger scale. Detailed descriptions of individual Soviet research institutes in the oil and gas sector show that much of what they do is production trouble shooting, routine economic analyses, simple geology, and so on. In the final analysis, however, the conclusion seems inescapable that with such large expenditures and the present level of Soviet fossil fuel technology, the R and D institutions in these conventional energy areas must display strongly the inefficiency thought to be common in much of Soviet R and D.

(6) Note that the geothermal and solar programs constitute even less palpable elements in the Soviet total than the United States program.

(7) Finally, note the large expenditures in the USSR for nonnuclear electric power. A curious feature of all the U.S. distributions is that they show very little R and D for electric power. The NSF breakdown of energy R and D performed in industry (see below) has no such category, except as it may be captured in "other," and in the NSF analysis of federal funds expenditures shown for electric power under the heading of conservation really *are* measures for "conservation" rather than for the development of electric power equipment. It seems to me possible that much R and D for developing new electric power equipment is not included in the U.S. totals. Consider the following NSF table on

R and D performed in industry for 1974, which shows expenditures by product field (in millions of dollars):

Engines and turbines	$406
Electrical equipment except communication	890
Electrical transmission and distribution	217
Electrical industrial apparatus	280
Other electrical equipment and supplies	393

All of the expenditures for developing electrical transmission and distribution equipment would seem to be energy-relevant, and there must be significant R and D for conventional power plant turbines in the "engines and turbines" category and for electric power generating equipment in the "other electrical equipment" category. But there is no room for expenditures of this magnitude in the alternative classification of R and D expenditures by areas, shown in the same source as follows (in millions of dollars):

Fossil fuel	$506
oil	336
gas	61
shale	10
coal	66
other	33
Nuclear	600
Geothermal	2
Solar	7
All other	82

R and D for transmission and distribution alone significantly exceeds the unidentified "other" categories where it would have to go. It may be that NSF wants its concept of energy R and D to exclude development of this kind of equipment, though it would seem that if the development of nuclear reactors is to be included, then improvement of conventional generating and transmission facilities should also be included. In any case, this difference helps explain why the Soviet total looks so large in relation to that for the United States.

But even with all these allowances, the final conclusion seems to be that, in the area of conventional energy sources and electric power generation, the USSR has such large expenditures that, in the light of the level of technology in those branches, it must be very inefficient in that kind of R and D. This is explained in terms of inertial growth and programming, reinventing the bicycle, duplication of effort, divorce of R and D from practical needs, and the other organizational pathologies often ascribed to Soviet R and D in general.

CONCLUSIONS

With this chapter, the task of Part I is completed. How energy policy is made has been described, as has the generally important role that innovation and technological change are called upon to perform as instruments of energy management. This chapter has shown that there have certainly been impressive amounts of resources devoted to the R and D activities that are intended to produce these innovations. One aspect of the matter not very carefully examined is whether this seemingly lavish provision of R and D resources for energy is a policy of long standing. In general, Soviet R and D employment and expenditures have grown during the post-World War II period at rates far above most economic subaggregates. Between 1950 and 1975, employment in Soviet R and D grew at a rate of 8.5 percent per year, compared to only 3.2 percent per year for industrial employment. Energy R and D growth has surely reflected this overall trend with reasonable sensitivity. Moved back to the fifties at something like the 8.5 percent rate cited, the amount of resources devoted to energy R and D in that decade was far smaller than today. But the inventory of these R and D institutions makes clear that many of them have existed for a long time and were large institutions already in the fifties. Whatever the history has been, however, enough R and D resources have been poured into solving energy problems that one is puzzled that they have not produced more rapid productivity growth.

Some hints of what may be the problem have been given along the way in the description of the R and D establishment, and some ideas of the weaknesses that might be expected to interfere with innovation in this kind of system have been offered in the earlier part of this chapter. But to obtain any concrete perspective on how well energy R and D operates, what problems may hamper its effectiveness, and how well it fits into overall energy management, it is necessary to look at actual experiences. What is required is a more detailed look at technological policies in particular sectors, what concrete R and D tasks have been, and how the planning, R and D performance, and mastering of specific innovations has worked. For this purpose the following chapters examine several examples of R and D and innovation at work. These chapters may be considered a kind of case study approach intended both to generate some judgments about the effectiveness of R and D and to reveal specific features of the Soviet approach to R and D that will be generalized in the final chapter.

3   Thermal Power Generation

It will be useful to have in mind some of the main technological facts and economic trade-offs in electric power as a basis for thinking about productivity change and relative technical levels. The concern in this chapter is primarily with fossil-fired thermal units; hydro power is more or less ignored, and nuclear power will be considered separately in Chapter 5.

Since the dominant inputs are capital, fuel, and (to a much smaller extent) labor, the strategic variables on which innovation efforts are concentrated are the heat rate (fuel expenditure per KWH generated) and the capital/output ratio. Reductions in the capital/output ratio are strongly related to increases in scale, at the level of the individual generating unit, at the level of the plant, and at the level of system integration. As the size of the individual generating unit increases, there are savings of metal per KW, smaller dimensions per KW and hence smaller requirements for buildings, and so on. Characteristic units today in the USSR are 300, 500, and most recently 800 MW blocks integrating a single boiler with a turbogenerator unit. Increasing size brings new problems such as cooling the generator (now solved by circulating hydrogen in it), and the problem of reliability grows as the number of stages and parts in a turbine increases. Since big units have high costs when they are only partially loaded, they must be used for base load. Further economies result from aggregating units into large plants, largely because of savings in the construction costs of the building and the auxiliary units that go along with the boiler-turbogenerator complex.

Fuel supply may place a limitation on the size of plant. A power plant fired with peat, which has low heating value per unit weight, faces high costs for collecting an adequate fuel supply; oil and gas are easier to supply in adequate quantities. The use of nuclear fuel eliminates this consideration, and indeed there are factors (such as security, reprocessing of fuel, and so on) that will probably encourage a shift to large complexes for nuclear plants. Another limitation on the size of thermal plants is the supply of cooling water. There is a serious shortage of water in the European USSR, and a limited number of suitable sites for big plants.

If the plant is to supply heat as well as power, plant size is constrained by the size and density of the heating load in the area. The size of the local power market is less important because it is possible to distribute power economically over much larger distances than heat. Heat and power turbines tend to be considerably smaller than condensing turbines—the largest Soviet heat and power turbine is a 250 MW unit.

One of the significant capital costs is for reserve capacity, both to meet peak demands and to cover equipment outages. The cost of reserve capacity can be reduced by optimizing peaking units in light of their low utilization—i.e., reducing their capital cost per KW at the expense of higher fuel and operating costs. A gas-turbine-powered unit has a comparatively high heat rate but has lower capital cost per KW of capacity. Other strategies for reducing the costs of reserve capacity are pumped water storage, allocation of hydro potential to this use, integration into networks, and increases in equipment reliability.

Fuel accounts for a very large fraction of production cost in thermal stations, and reducing the heat rate is a central preoccupation. In the outlays of Soviet power stations, fuel typically accounts for more than 50 percent. That fuel is a relatively expensive input in the Soviet case is demonstrated by the following statistics. The average delivered cost of fuel to thermal stations in the USSR circa 1974 was 15.5 rubles per ton of standard fuel (Avrukh, 1977, p. 149). The cost of fuel has been near this level for a long time—it was 14.28 rubles per ton in 1950, then fell gradually to about 11 rubles per ton in the sixties (Avrukh, 1966, p. 125). The rise to the present level was caused by the 1967 price reform. The average in the United States was about 8.5 dollars per ton of standard fuel in 1970 and 25 dollars per ton of standard fuel in 1974 (FPC, *Steam-Electric Plant Construction Cost and Annual Production Expenses*, 1974). This would make the ruble/dollar ratio about 1.8 in 1970 and about 0.6 in 1974. The corresponding ratio for Soviet to U.S. average wage and Soviet to U.S. cost per unit of capacity were respectively 0.16 and 0.5. These are approximate but reasonably accurate ruble/dollar ratios based on standard sources for wages (FPC, op. cit.) for U.S. capital costs, and on a Soviet figure of 100 rubles per

KW, which is suggested in many sources.* As the comparison suggests, the cost of fuel relative to other inputs has in the USSR long been at the level that has emerged in the United States only as a result of the post-1973 adjustments in world energy prices. This helps explain the great importance reducing the heat rate has had in Soviet policy. The major route to that end is by raising steam pressure and temperature. The progression in the USSR has been from steam at 35 atmospheres before World War II to 90 atmospheres after the war, then to 130 atmospheres and now 240 atmospheres. Temperatures have risen correspondingly, and the standard temperature is now 540°C. Most of the new units thus operate at supercritical steam conditions—i.e., above 221 atmospheres and 374°C.

Higher temperatures and pressures involve more expensive equipment, and the relevant trade-off is fuel savings versus the extra investment. Thus, there has been a long controversy as to whether the proposed complexes of generating capacity based on the Kansk-Achinsk and Ekibastuz coal deposits (where fuel is very cheap) justify the use of supercritical steam parameters with the associated higher capital costs.

The thermodynamic properties of power generation can also be improved by combined cycles. Two main approaches are relevant to the Soviet case. Fuel may first be utilized at high temperatures in gas turbines or in magnetohydrodynamic generators, with the exhaust from these devices used to generate steam for conventional turbines. In power and heat combines, the remaining heat of the steam after partial expansion in turbines is captured for heating purposes rather than being rejected to the environment.

Fuel economies are also obtainable through improved combustion and through shifts from one fuel to another. Gas and oil are more efficiently burned than coal and other low grade fuels.

Fuel expenditures are influenced by load characteristics as well. A fixed fuel cost is incurred in keeping a unit spinning on the line, and the fuel input per unit of power generated drops as the unit becomes fully loaded. So integration of systems, effectiveness of load dispatching, and all the innovations that underlie them are important in reducing fuel input.

Labor in a power plant is more involved in servicing the equipment

---

*It is quite possible that this is understated, and if some of the highest cost figures associated with obviously more capital-intensive uses such as nuclear are used—say up to 140 R/KW—the ruble/dollar ratio would rise to 0.7.

than in handling inputs and outputs, so labor costs are more a function of repair needs, degree of automation of control, and so on than of amount produced. There are large economies of scale here—the bigger the unit and the bigger the plant, the smaller the labor input per unit of capacity being served. Probably the most important determinant of labor requirements is the degree of automation. The size of the plant that can be handled is a function of the sophistication of control equipment, and I suspect that for a long time the Russians were backward in this area. Certainly great attention is given to it in their discussions of technological progress achieved and still to be accomplished. (There is an especially informative discussion of this matter in Shvets 1975, pp. 78–89).

A number of dimensions of output are seen as constraints rather than as variables in the Soviet approach to electric power decision making. Current is supposed to meet frequency, voltage, and continuous availability norms; though these are often violated, this is the result of poor planning rather than of an optimizing calculation. Environmental effects are also often seen as constraints rather than choice variables in system optimization. Soviet policymakers may be only loosely constrained on environmental effects but they do not usually treat them as variables to be traded off against other desiderata. They may pollute the air, not because they consciously conclude that this is better than the expense of amelioration measures, but because they ignore clean air as a desideratum, unless some outside source makes them consider it.

TECHNOLOGICAL LEVEL AND PROGRESS IN SOVIET POWER GENERATION

Aggregate data for major inputs and outputs in the United States and the Soviet Union for the electric utility sector (shown in Table 3-1) indicate high comparative productivity and, by inference, a comparatively high technical level for the USSR in this branch. The Soviet industry has higher utilization of installed capacity and a lower heat rate. Soviet labor productivity is only 29 percent of the U.S. level, but, considering that in coal mining Soviet labor productivity is only 7 percent of U.S. level, this must be considered a comparatively good showing. If we use the 0.5–0.7 range for the ruble/dollar ratio for electric power generating capacity suggested earlier, to put a dollar cost on the resources embodied in the Soviet capital stock, the USSR uses a stock

TABLE 3-1. Comparative Indicators for U.S. and Soviet Electric Power
Utilities, 1975

		USSR	U.S.	U.S./USSR
Output	(GWH)	962[a]	1,917	2.00
Thermal	"	836	1,617	1.93
Fossil-fired	"	817	1,445	1.77
Employment (thousands)[b]		686	503	0.73
Employment		—	401	0.58
Capital				
Value of fixed assets (billions)		54 rubles	189 dollars	
Average annual capacity (MKW)		187	491	2.62
Fuel consumed in fossil-fired thermal (MT of standard fuel)[c]		281	540	1.92

SOURCES AND NOTES: Soviet data from standard Soviet statistical sources. U.S.
data from Edison Electric Institute, *Statistical Yearbook, FPC, Statistics of Publicly
Owned Utilities in the U.S., Statistical Abstract of the U.S.*

[a]These Soviet output figures differ from those reported in standard Soviet sources
gross of station use. Correction to the net generation concept used in the United
States is based on Nekrasov and Pervukhin, 1977, p. 18.

[b]The Soviet figure on employment refers to *promyshlenno-proizvodstvennyi personal'*, that is, it excludes employees in ancillary activities such as construction.
The larger figure for the United States is the sum of employment in privately
owned utilities (Edison Electric Institute) and publicly owned utilities. For the
latter, the U.S. *Statistical Abstract* gives a figure for municipal utilities which I
have adjusted upward for federal utilities in proportion to generation. The 503
thousand total includes 101 thousand construction account workers in private utilities, and these have been removed in the smaller figure. On the other hand, the
Soviet figure probably contains a large number of personnel serving the by-product
heating function characteristic of many Soviet power stations.

[c]To convert from BTU to the Soviet measure of tons of standard fuel, one quad
(one quadrillion BTUs) $= 36 \times 10^6$ tons of standard fuel.

57–60 percent as large as ours to produce an output half as big. That
seems a relatively good achievement and is all the more impressive
when it is realized that the USSR has a higher ratio of transmission
line to generating capacity than the United States does, and a higher
share of its generating capacity in the very capital-intensive hydroelectric branch. Two of these indicators—the heat rate and utilization
of capacity—merit more detailed discussion.

The Heat Rate

Comparisons of the heat rate over time with the U.S. and Western
Europe are shown in Figure 3-1. In the early postwar period the USSR

FIGURE 3–1. Heat Rate in Thermal Electric Power Generation, Utility
Sector, United States, USSR, and Western Europe
(grams of standard fuel per KWH)

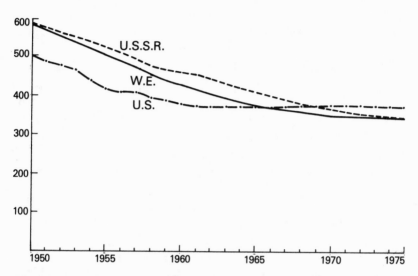

SOURCES: U.S. data from Edison Institute, *Historical Statistics of the Electric Utility
Industry*, and *Statistical Yearbook*. Soviet data from standard Soviet statistical handbooks. Western Europe here means the six original nations of the Common Market
and is taken from EEC, *Energy Statistics Yearbook*. Later expansion to nine mem-
bers raised the group average slightly, but the trend remained the same.

reported a heat rate for utility stations 20 percent above the rate for
the United States. But the rate has come down much more rapidly in
the USSR than in the United States and the USSR now claims an
average rate appreciably below the United States. The USSR has also
caught up with the heat rate achieved in Western European countries.

The Soviet heat rate is figured on a slightly different basis than in
the United States. The U.S. method uses the "higher heating value"
for fuel, including the latent heat of water vapor in the combustion
products, while the standard fuel in which the Russians give heat rates
is defined in terms of "lower heating value" (FPC, *Steam Electric
Plant Construction Cost and Annual Production Expenses*; and Chernukhin and Flakserman, 1975, p. 290). Adjustment of the Soviet rate
for 1975 to the American standard would raise it to 362 instead of 340,
but would still leave it below the U.S. rate (based on data in Ravich,
1977, p. 27). But even with this correction, the Soviet heat rate still

seems suspect to me. Certain differences give the Russians an advantage in fuel economy. First, there is a fairly heavy emphasis on heat and power combines (see below). Secondly, the share of supercritical units is probably slightly higher in total Soviet capacity than in U.S. capacity because of the difference in growth dynamics.

On the other hand, the low quality of the fuel burned in Soviet plants should raise fuel expenditure appreciably. The average heat content of coal burned in Soviet electric power plants is less than 4 gigacalories per ton (*Elektricheskie stantsii*, 1974:4, p. 13), compared to a U.S. average over 6 gigacalories per ton. Soviet authorities themselves say that this kind of fuel places a heavy constraint on the efficiency of combustion. One source cites a conversion from Cheliabinsk lignite to gas that raised boiler efficiency from 88 to 94 percent, and one from anthracite to gas that raised efficiency from 91 to 95 percent. (Ravich, 1974, p. 234.) It is difficult for me to believe that Soviet fuel performance for all condensing stations (reported as 365 grams of standard fuel per KWH in 1975—*Elektricheskie stantsii*, 1977:1, p. 82— or 389 grams in higher heating value) is that close to the U.S. average for all stations, reported as 374 grams/KWH in 1975.

In the USSR there must be a strong temptation to report good fuel performance and therefore a possible bias in the reporting system. According to Pruzner (1969), the heat rate is not part of the official plan of either the system or the plant but is figured as an auxiliary indicator. But this rate has to be important at some level and the Gosplan's planning manual in its section on planning the power sector states that "besides the general economic indicators, among the most important indicators of effectiveness, the improvement of which must be envisaged in the plan, are the heat rate and the expenditure of fuel per gigacalorie of heat . . ." (Gosplan SSSR, 1974, p. 86). One hint that Soviet power plants may underestimate fuel consumption is found in a study advocating pricing energy coal according to heat value (Ivasenko, 1977). The author cites average heat contents reported separately by mines and by the receiving power stations for 95 lots of coal totalling 104 thousand tons. In 79 cases the power plants reported lower calorific values; in 21 cases they were more than 800 KKal/Kg lower, in another 27 cases 300–800 KKal/Kg lower (page 68). The mine may also have exaggerated heat content, but it seems to me its motive for doing so is less strong than the power plant's for understating it.

Utilization of Installed Capacity

The Soviet figure for average hours of utilization of installed capacity has consistently been above that for the United States. (It was 5,377 hours for the USSR versus 4,332 hours for the United States in 1960, with the difference gradually increasing in recent years to 5,257 hours and 3,906 hours, respectively, in 1975.) A higher rate of utilization can reflect such aspects of technological progress as system integration, reduction in time lost through breakdowns and repair, and improvements in scheduling and locating capacity additions in relation to demand growth. In this case, however, most of it is caused by differences in load conditions, and by a Soviet willingness to do without reserves at peak demand.

The difference in average annual hours of utilization can be accounted for by two factors. First, the USSR has little reserve capacity. The ratio of peak demand to installed capacity in the power system of the Center region in 1970 was an incredible 0.98, and, though smaller elsewhere, it is still high overall. This is probably a weakness rather than a technological or planning accomplishment, since it often leads to cutting consumers off at peak load periods, and not maintaining frequency and voltage standards. Secondly, the USSR enjoys flatter load curves. The load factors for various Soviet systems, (figured on an annual basis by comparing annual output with peak load times the number of hours in the year) tend to be 0.65–0.69, whereas in the United States they are more likely to be 0.57–0.65. An average for the contiguous United States would be about 0.62 and for the European USSR about 0.68. This factor alone accounts for most of the difference in average hours of utilization. Correction for both these factors would make Soviet utilization of installed capacity inferior to U.S. performance.

The higher performance of Soviet power generation as measured by input productivity is mostly explainable by success in developing reasonably large and efficient generating equipment. Significant economies come from increased size of turbogenerator units and plants and from improving steam conditions. Moreover these achievements reflect numerous dimensions of technological sophistication and input quality such as steel quality, construction techniques, automation of station control, and so on. In the remainder of this chapter, therefore, discussion will focus on several aspects of Soviet experience in creating the basic equipment for fossil fired turbogeneration of electricity. This experience is also useful in revealing distinctive features and specific strengths and weaknesses in the Soviet approach to R and D.

CHARACTERISTICS OF SOVIET POWER GENERATING EQUIPMENT

The changing structure of fossil-fired generating equipment by size of units, by steam parameters, and by size of station is shown in Tables 3-2, 3-3, and 3-4. Consider first the size of individual units.

The Soviet distribution by unit size has generally been somewhat inferior to those of the United States and Western Europe, as the USSR has lagged consistently behind the Western countries in introducing each successive generation of larger generating units. As of

TABLE 3-2. Distribution of Steam Turbogenerator Units by Size of Unit, Regional or Utility Stations, End of Year, Percent of Total Steam Turbine Capacity

Size of Unit (MW)	1950	1955	1960	1965	1970	1974	1975 Plan
800	0	0	0	0	0.6	0.5	1.95
500	0	0	0	0	0.4	0.3	0.90
300	0	0	0	4.4	16.8	24.0	22.60
200[a]	0	0	0.4	11.9	13.5	c	14.60
150[b]	0	1.7	2.3	11.8	10.5	24.5[c]	14.1
100–149	0	0	22.0	12.0	11.9	c	8.65
less than 100	100	98.7	75.3	59.9	46.3	50.6	37.2
50–99	7.5	18.4	21.1	n.a.	n.a.	n.a.	n.a.
24–50	55.2	55.9	36.9	n.a.	n.a.	n.a.	n.a.
12–24	20.3	13.2	9.3	n.a.	n.a.	n.a.	n.a.
6–12	9.9	6.4	4.0	n.a.	n.a.	n.a.	n.a.
less than 6	7.1	4.4	4.0	n.a.	n.a.	n.a.	n.a.
Total capacity of units covered (MW)	n.a.	n.a.	n.a.	80.9	123.1	155.0	n.a.

SOURCES: 1950–1960: Vilenskii, 1969, pp. 147–151, except that distribution of 150 MW and above for 1955 and 1960 is based on Pavlenko and Nekrasov, 1972, p. 102. For 1965, 1970, and 1975 plan the source is Pavlenko and Nekrasov, 1972, pp. 108, 110. 1974 is from *Elektricheskie stantsii*, 1975:1.
[a]includes for 1975 Plan the 250 MW heat and power turbine.
[b]includes a couple of 160 and 170 MW units.
[c]24.5 is the total for these three classes.

1974 (the last year for which the actual Soviet structure seems to have been disclosed), the comparison with Western Europe (specifically the nine Common Market countries) shows the Russians with still over half their capacity in the less-than-100 MW category, compared to only 29.3 percent for the Western European group, and with only 0.8 percent in the 500 MW-and-up class, compared to 20.1 percent in Western Europe (EEC, *Energy Statistics Yearbook*). One reason for the relatively high share of equipment with capacities less than 100 MW is the heavy use of heat and power equipment in the USSR, a peculiarity of the Soviet structure that has compensating advantages. It seems impossible to make the same detailed comparison with the United States, but it is possible to estimate the share of equipment in the 500 MW-and-up and the 300MW-and-up classes by cumulating information on new additions in FPC, *Steam Electric Plant Construction Costs and Annual Production Expenses*. This shows that at the end of 1974, the United States had about 49 percent of its fossil-fired steam capacity in units with capacities of 300 MW and over (compared to 24.8 percent in the USSR) and 27.5 percent in the 500 MW-and-over class (compared to 0.8 percent in the USSR).

The share of units with higher steam parameters has grown significantly, as shown in Table 3-3. In particular, the share of units using steam at supercritical conditions rose in the decade from 1965 to 1975

TABLE 3-3. Steam Parameters of Thermal Power Generating Units, End of Year, Condensing and Heat and Power Stations

Pressure (atm)	1958 (GW)	percent	1965 (GW)	percent	1970 (GW)	percent	1975 Plan (GW)	percent
300	0	0	0	0	0.1	0.08	0.10	0.06
240	0	0	3.60	4.45	22.0	17.85	43.40	26.35
200	0	0	0.05	0.06	0.05	0.04	0.05	0.03
170	0.66	1.9	0.60	0.75	0.60	0.50	0.60	0.37
130	0.20	0.6	28.26	34.90	47.20	38.40	67.22	40.90
60-120	14.85[b]	43.8[b]	2.70	3.34	2.70	2.20	2.70	1.60
90			24.25	30.00	29.90	24.30	32.67	19.85
less than 35	18.19[a]	53.7	21.40	26.50	20.52	16.63	17.80	10.84
Total	33.90	100.0	80.86	100.00	123.07	100.00	164.54	100.00

SOURCES: 1958: *Energetik*, 1966:8, p. 2. Other years: Pavlenko and Nekrasov, 1972, pp. 94–95, 108, 110.

[a]includes some units up to 76 atmospheres, but most are in units of 35 atmospheres or less.

[b]includes 60–120.

TABLE 3-4. Distribution of Installed Capacity by Capacity of Station, End of Year, Percent of Total

Class (MW)	1950	1955	1958	1960	1965	1967*	1970	1975
up to 0.5	12.8	13.5	15.2	14.2	11.7	n.a.	n.a.	n.a.
0.5– 9.999	17.1	11.8	8.9	7.9	5.7	n.a.	n.a.	n.a.
10–49.999	18.8	13.9	10.7	8.6	6.8	n.a.	n.a.	n.a.
50–99.999	14.2	13.2	9.3	8.0	5.4	n.a.	n.a.	n.a.
100+	37.1	47.6	55.9	61.3	70.4	74.1	n.a.	n.a.
100–200	n.a.	n.a.	n.a.	n.a.	n.a.	10.2	n.a.	n.a.
200–400	n.a.	n.a.	n.a.	n.a.	n.a.	15.3	n.a.	n.a.
400–600	n.a.	n.a.	n.a.	n.a.	n.a.	9.3	n.a.	n.a.
600–1,000	n.a.	n.a.	n.a.	n.a.	n.a.	14.4	n.a.	n.a.
1,000+	0.0	0.0	0.0	2.3	~18.0	24.9	40.7	48.1
2,000+	0.0	0.0	0.0	0.0	2.3	n.a.	16.1	27.8
Total capacity (MW)	19.614	37.246	53.641	66.721	115.033	131.727	166.150	217.484

*On the reasonable assumption that all stations above 100 MW are utility stations.

SOURCES: 1950–1967: Vilenskii, 1969, pp. 130–31, 243, except 1,000 MW and above for 1965, from *Elektricheskie stantsii*, 1975:4. 1970: *Elektricheskie stantsii*, 1975:4. 1975: *Energetik*, 1976:6, inside front cover.

from only 4 percent to about a quarter of the total. (I imagine that the planned 1975 distribution was approximately achieved.) This is probably one of the major factors in the good showing on heat rate performance compared to the United States, where the share of supercritical units in total capacity in 1974 was only about 20 percent.*

Parallel with the growth in unit size there has occurred over the same period a dramatic growth in station size, as shown in Table 3-4. (Note that this table refers to *all* utility stations, including hydro and nuclear.) In 1955 the largest station had a capacity of 510 MW, and in 1960 there was but one station with a capacity in excess of 1,000 MW. In the sixties a new strategy of building large stations was adopted, and in 1975 there were 43 stations with capacities exceeding 1,000 MW. The maximum station capacity at that date was 3,000 MW, but, with introduction of the 500 and 800 MW units, stations as large as

*There were 123 such units in operation in the United States at the end of of 1974. It is not worth the effort to reconstruct their total capacity from the plant data in FPC, *Steam Electric Plant Construction Costs and Annual Production Expenses*, but the average unit capacity cannot have been far from 600 MW, which would make their share in all steam units 20.1 percent.

5,000 MW are expected to be built in the future. The growth in the larger stations comes as a result of both the building of new stations and the enlargement of older ones. For example, the Krivoi Rog regional station, by adding 300 MW units, has grown from 1,800 MW at the beginning of 1970, to 2,400 MW by the beginning of 1971, 2,700 MW by the end of 1972, and 3,000 MW at the beginning of 1974.

Distributions fully comparable to Table 3-4 are not available for Western Europe and the United States, but it appears that the Soviet distribution is distinguished both by a large legacy of very small plants, and by a tendency to build very large plants. In 1965, for instance, when the U.S. private utility sector contained no stations as small as 10 MW, the USSR had 17.4 percent of its capacity in that category (FPC, *Statistics of Privately Owned Electric Utilities*). Publicly owned utilities in the United States probably had smaller plants on the average than privately owned utilities, but their inclusion would probably not affect this particular comparison. At the same time the USSR had 18 percent of its capacity in stations with capacities of 1 GW and over, whereas in the United States only 12 percent of the capacity of utility stations was in units this large. The trend to large stations continued apace in the USSR in the decade following 1965, though the comparative standing in the share of 1 GW and over was about the same in 1975, when the United States had 31 percent of its capacity in such plants compared to 48.1 percent in the USSR. In the distribution for the United States, plants with capacities of 2 GW and up are not even distinguished, but it is possible to identify them fairly easily in the FPC plant data (FPC, *Steam Electric Plant Construction Costs and Annual Production Expenses*, and *Hydroelectric Plant Construction Costs and Annual Production Expenses*). They accounted for 10.3 percent in the United States and 27.8 percent in the USSR in 1975. In comparison with Western Europe also, the USSR has a heavy emphasis on large plants—in the nine Common Market countries at the end of 1975, stations with capacities 1 GW and up accounted for 41.1 percent and plants of 2 GW and up for 17.4 percent of all thermal plants (EEC, *Energy Statistics Yearbook*—this tabulation excluded hydro plants, but their inclusion could not change the picture much).

This background makes obvious the importance of improved equipment as the main technical line for raising the technological level of power generation. I shall accordingly devote the rest of this chapter to examining the development histories of some of the main models of generating equipment. The growth in station size is less obviously a matter of technological sophistication as such, and I will have much less to say about it.

BASE-LOAD BLOCKS FOR CONDENSING STATIONS

The principal steps in the history of development of equipment for fossil-fired condensing plants can be briefly summarized. In the prewar period the basic units in condensing stations operated at 35 atmospheres, but after the second World War a shift was made to units at 90 atmospheres and 500°C as the mainstay in new additions. Units with 150 and 200 MW capacity with steam at 130 atmospheres and 565°C were first introduced at the end of the fifties. The first K-150-130 was installed in 1958, and the first K-200-130 in 1960 (Neporozhnyi, 1970, p. 203). These were also the first standard units to embody block design—i.e., units in which the boiler, turbine, generator, and transformer are combined in an integral unit. Block design decreases the capital expenditure per KW of capacity by decreasing the cost of components and the volume of building necessary to house the equipment. The 150 and 200 MW units constituted the main additions during the sixties—some 141 units were installed during the decade, and a significant number of them are still being installed today.

During the years these units were being developed the producers were experimenting with still higher steam parameters. In 1952, a 150 MW unit at 170 atmospheres and 520°C was installed in the Cherepovets station. The Leningrad Metal Factory (LMZ) produced a 50 MW unit at 200 atmospheres and 400°C to add onto an existing turbine at the Cheliabinsk station, and a unit at 300 atmospheres and 650°C with reheat, designed and produced by the Khar'kov Turbine Factory (KhTZ), was added onto an existing turbine at the Kashira plant. All used austenitic steel, but it was decided that this was too expensive and that future designs would use only perlitic steel. At the end of the fifties a decision was being made as to what the next step should be to produce a standard unit at supercritical parameters. One opinion was that it would be possible by 1965 to produce a 600 MW unit, at 580° and 240 atmospheres, using perlitic steel (*Elektricheskie stantsii*, 1958:2, pp. 2–6 and 1959:7, p. 2). But for some reason it was decided that this was too bold a step, and further tests also led the designers to conclude that given the quality of available perlitic steel, the design temperature should be 565°C rather than 580°C. One source suggests that the ambitious designs were favored as the result of a kind of "engineering prejudice" favoring a lower heat rate as an end in itself, and that the decision for a less ambitious design was a victory for those who in the second half of the fifties helped strengthen economic calculation and sophistication in the power industry (Kulikov, 1964, pp. 122–124). The upshot was a decision for a 300 MW unit at supercritical steam conditions of 240 atmospheres and 565°C for both initial and reheat temperatures. The history of this model is instructive in assessing the effectiveness of R and D in this field, and worth describing in detail. LMZ and KhTZ each produced a prototype (what the Russians call a *golovnoi obrazets*) in 1960–61 (*Vestnik mashinostroeniia*, 1967:10, pp. 18–22). These were installed by the end of 1963, at the Cherepovets station, and the Pridniepr station (Pavlenko and Nekrasov, 1972, p. 100). As background, consider that the United States had installed the first 500 MW unit at supercritical temperature and pressure in 1960, and in 1965 the first 1,000 MW unit (*Electrical World*, March 4, 1968, p. 34; and CIA, 1965, p. 6). This became the basic unit for new condensing stations. Three more of the 300 MW units were installed in the USSR in 1964 and five in 1965. Annual numbers built up to a peak of 15 or 16 in the early seventies and then gradually fell off. Some will still be installed in the 10th Five Year Plan.

The most striking aspect of introducing the 300 MW model is how long it took to "master" these units once they had been installed. Typically only after four or five years would a unit operate anywhere near its design load or its design fuel rate. In 1965, when 12 were in operation, their *average* availability was only 41.2 percent, and the heat rate was 416 grams/KWH versus the design rate of about 320 grams. Problems were multiplied when the fuel presented any unusual difficulties. The Ermakov plant burning Ekibastuz coal had the following indicators:

	1969	1970	1971	1972	1973	1974
utilization (percent)	16.9	33.4	40.0	53.0	62.2	73.2
heat rate (grams/KWH)	613	448	392	364	346	340

(*Elektricheskie stantsii*, 1975:3)

Thus, it was only in the *sixth* year of operation that it achieved tolerably satisfactory operation. There were frequent breakdowns among the 300 MW units, 60 percent of which were due to damage to the heating surfaces of the boilers (*Energetik*, 1975:2, p. 4). But reliability problems arose with all major components, including turbines, condensers, feed pumps, and hot parts. Somewhere during the process it was found that steel quality was inadequate for operation at 565°C, leading to excessive breakdowns, and a decision was made to operate *all* equipment at 545°C both for initial steam and for reheat. Since Soviet commentators do not like to talk about this, there is not much information on when and how this decision was reached, but it is stated unambiguously to be the case (Leonkov, 1974, p. 74).

One interpretation for slow mastery and poor performance of the 300 MW units is that "serial production of the 300 MW units was begun before adequate experience had been gained by operating the prototype. Some of the first 300 MW machines installed may never operate satisfactorily . . ." (CIA, 1965, p. 6). As time went on, experience with getting these sets to operate did improve—it is said that those installed in the early seventies could be brought somewhere close to their design indicators within a few months rather than within a few years.

Another difficulty with the protoypes and early series units was failure to design and deliver suitable auxiliary equipment along with the main units. In the early stages it was necessary in installing these blocks to use makeshift equipment developed for other purposes for pumping, draft, deaeration, feed water heating, and other tasks (*Elektricheskie stantsii*, 1968:8, pp. 17–18). This is a traditional problem, as the following description of experience with the earlier 160–200 MW units shows:

> The principal boiler and turbine factories do not provide for complete delivery of auxiliary equipment. . . . For example, while new types of boilers with steam productivity of 500–640 t/hr already are being manufactured by factories not even the drawings have been produced for the auxiliary equipment—feed pumps, coal pulverizers, draft equipment, ash removal equipment. . . . Incomplete deliveries take place also with respect to turbines. Currently the first turbines of 150–200 MW capacity are already being manufactured, while the necessary condenser pumps, vaporizers and various types of fittings for them are lacking. [*Elektricheskie stantsii*, 1958:5]

This is more than just a failure of coordination in deliveries, and reflects rather a failure to coordinate and phase correctly the various parts of the development program. Indeed, the history of the 300 MW units suggests that there is something even more fundamentally wrong.

The original conception was never realized in a perfected complex of equipment. The average heat rate for the 300 MW units in 1975 was still 341 grams/KWH rather than the design figure of 320 grams. The lowest rate I have seen quoted for *any* single unit is about 334 grams/ KWH for mazut-fired stations and 355 grams/KWH for coal-fired ones (*Energetik*, 1973:3, p. 4). This is not much better than what is achieved in the 200 MW units at 130 atmospheres. Moreover, the supposed advantage of these larger units—savings in capital costs—has apparently not been realized. Whereas it was expected that the 300 MW unit would be cheaper per unit of capacity than the 200 MW units, it is in fact more expensive (Karnaev, 1972, p. 105). This 300 MW unit has worked so badly that the 160 MW units and 200 MW units have continued to be produced in significant numbers in the seventies (Pavlenko and Nekrasov, 1972, p. 102), though it was being said in 1968 that in the next couple of years production of the 160 MW unit would be completely stopped (*Elektricheskie stantsii*, 1968:8, pp. 17–18).

Experience with the next step upward to 500 MW and 800 MW block units has been roughly similar. The specifications (*tekhnicheskoe zadanie*, the documents guiding the design of a new unit) for these were approved by 1962, and it was planned that mastering (*vnedrenie*, which I take to mean testing after installation) would begin in 1965 (*Elektricheskie stantsii*, 1962:8, p. 3). There was some slippage and the first 800 MW unit in a two-shaft version did not actually begin operation until 1968 (Pavlenko and Nekrasov, 1972, p. 100). But this model "did not justify itself in operation" and no further such units were produced (*Energetik*, 1975:1, pp. 5–6). I do not know all that was wrong with this prototype, but one of the main difficulties was with the boiler; it had been designed to work on coal, but this apparently was impossible, and in the end the boiler was totally redesigned to work on gas and residual fuel oil (*mazut*), then finally on *mazut* alone (*Energetik*, 1975:3, p. 7; and *Elektricheskie stantsii*, 1975:1). Six years after installation, in 1974, it actually began to work decently, producing 5.1 BKWH, but its heat rate was still 353 grams/KWH (*Energetik*, 1975:3, p. 7). The second 800 MW unit (the one-shaft version) was "accepted for operation" in December 1971 and then worked badly for the next two years. Only sometime in 1973 did it get to full power (*Energetik*, 1974:1, pp. 5–6). In 1974, it achieved more or less normal operation, producing 4.25 billion KWH (61 percent utilization of capacity) at a heat rate of 366 grams/KWH. I have not seen what is

claimed as the design heat rate for these units, but since it was supposed to be somewhat below the design rate for the 300 MW units
(320 grams/KWH), 366 grams is far above the design rate.

The next two 800 MW units were commissioned sometime during
1975, at the Uglegorsk and Zaporozh'e stations. They seem to have
been mastered more rapidly. In 1976, the coefficient of use of capacity
was only 49.2 percent, and the heat rate very high (*Energetik*, 1976:9,
p. 2). Another source, however, says they have carried a load at nominal capacity and have had periods of operation at a heat rate of 322–
328 grams/KWH (*Energetik*, 1977:4, p. 6).

The first 500 MW prototype was installed at Nazarovo. Originally,
vnedrenie was to begin in 1965, though in fact the station was not
commissioned until 1968.* A second 500 MW unit was installed at the
Troitsk station and commissioned 24 June 1974. According to Pavlenko and Nekrasov (1972, p. 233), it was supposed to have been commissioned in 1973. The Nazarovo and Troitsk models were alternative
designs intended to constitute a kind of design competition. The one
at Nazarovo apparently never worked at all, and so it is impossible to
find much detailed discussion about what the problems were. The
other, burning Ekibastuz coal, was made to work after many adjustments and modifications, and an improved second prototype was installed in the same plant. One source says that with this second unit
the problem of burning Ekibastuz coal in a 500 MW block has been
mastered.

One puzzle about these larger units is that small reductions in heat
rate seem to have been bought at very high costs in extra investment.
It is said that in the 800 MW turbines the cost per KW is double that
for the 300 MW units (*Planovoe khoziaistvo*, 1976:10, p. 11). Of
course, there may be some offsetting gains in other elements of the
whole complex.

It is intended that during the 10th Five Year Plan the new 500 and
800 MW units will be produced and installed on a considerable scale
in fossil-fired condensing stations as the standard models (*Planovoe
khoziaistvo*, 1976:3, p. 63). They are supposed to be supplemented
with a newly developed 1200 MW unit, 300 MW units, and a new
500 MW semipeaking unit, to be discussed later (Nekrasov and Per-

---

*This is the unit which, according to Hedrick Smith in *The Russians*, was commissioned in 1968, amid stories in the press about the surge of power into the
Siberian grid, when in fact the *generator* was not yet even on the site! Five years
later the unit was still not in operation.

vukhin, 1977, p. 222). But if things go as they usually have in the past, it is likely that all of the newer models will be delayed so that the 300 MW units will play a larger continuing role than intended.

This brief description seems to support several conclusions. The Soviet Union is still perhaps a decade behind the Western countries in its ability to design and produce large scale conventional steam-based power sets. Each step upward is one in which the experience of the Western countries has already shown what is feasible, but even to follow along behind has strained Soviet R and D capabilities to the limit. There seems to be a poor progression through successive R and D stages, with a tendency to move ahead fairly early to a prototype that is probably inadequately supported by previous experience, design, and material studies. The Soviet system seems to follow a philosophy of getting the new designs into production and then patching them up as production, installation, and operation proceed.

HEAT AND POWER COMBINES

One of the distinctive features of Soviet energy policy is heavy reliance on heat and power combines (*teploelektrotsentraly* or TETs), which in 1975 generated almost exactly one-third of all Soviet fossil-fired power output. Such combines offer the possibility of large fuel savings through capture of what would otherwise be rejected heat. But the intensity of the Soviet commitment to this idea has the smell of an engineering prejudice, and it deserves careful examination. As an indication of the commitment to this principle, it is universally accepted that future nuclear plants should be designed as heat and power combines. That means they would have to be located in heavily populated areas. Also, since nuclear plants are going to be very large (the Leningrad plant will have a capacity of 2 gigawatts) the heat consuming area and hence the average transport distance for steam would have to be relatively large. It seems possible this is one of those cases in which a principle is established at the strategic level, without taking into account the operating realities at the lower levels, which lead in operation to results much less favorable than envisaged by the planners. The Soviet power industry does obtain significant fuel savings from *teplofikatsiia* (as this practice is called), but its potential has been eroded by improvements in the equipment available for condensing stations, changing fuel prices, and mistaken decisions in the design of the dual purpose equipment.

The statistical basis for evaluating the fuel savings from TETs is tricky, since there are some heat and power units in what are basically condensing stations, and since in both TETs and condensing stations some heat is supplied to the heating network in the form of steam and hot water produced in separate units. Nevertheless, the situation is approximately as follows: in 1970, the pure TETs had a heat rate of 325 grams of standard fuel per KWH, whereas the average heat rate for condensing thermal stations (KES) was 389 grams (*Elektricheskie stantsii*, 1977:1, p. 82). Since the TETs produced about 250 BKWH of electric power in 1970, the saving would be about 16 million tons of standard fuel. This is a little misleading, however, since the heat rate cited for KES includes many old stations. The rate for the more modern equipment (i.e., all units using equipment at 130 atmospheres and up) was 368 grams/KWH, and, measured against this standard, the saving would be only about 11 million tons of standard fuel. The saving has grown over time, both because TETs output increases and because the heat rate of the combines has fallen relative to condensing stations. By 1975, the saving would have grown to about 25 MT of standard fuel (ibid.). These savings are large in absolute amount, but look rather small when measured against the total fuel consumption by all stations of 472.4 MT of standard fuel in 1974 and in view of the high share of TETs in all output. Soviet authors often cite much bigger figures for the fuel savings from TETs on the argument that, if this heat were supplied by individual units, the latter would be burdened with very low thermal efficiencies. But those gains are equally available from centralizing heat production alone, and should probably not be counted as gains from *teplofikatsiia* as a technological principle.

The explanation for this surprisingly low saving is that the TETs work a great deal of the time as condensing stations. In addition most are small and have unfavorable steam parameters. At the end of 1970, there was no cogeneration equipment with supercritical parameters, and units at 130 atmospheres and 565°C accounted for only 15 percent of all TETs' capacity (Pavlenko and Nekrasov, 1972, p. 94). By 1975, a few supercritical extraction turbines were in operation. Because of the relatively small size of these units, nonfuel costs (such as labor, amortization, and if it were counted, interest) are also relatively high for electric power produced in TETs, and the fuel saving is bought at a considerable cost in other resources. An authoritative work on power station costs says that investment per unit of capacity is 60 percent higher for TETs than for KES (Avrukh, 1977, p. 23). Another relevant consideration is that heat and power turbines get less intensive utilization as power producers than do condensing units—average annual use is 5,300 hours for TETs vs. 5,432 hours for all thermal stations in 1970 (*Elektricheskie stantsii*, 1974:11, p. 7).

The most important reason the stations have not lived up to their promise for fuel saving is that the heating potential their designers planned for has been underutilized. Most Soviet heat and power turbines are equipped with condensers (about 90 percent of all heat and power turbines by capacity), and when there is no heat load they work as pure condensing turbines. One source says, "As is well known, the heat rate for power in a large number of TETs and in TETs in general is significantly above the projected rate, which is explained basically by the underutilization of their heat capacity" (*Elektricheskie stantsii*, 1977:8, p. 17). The fraction of operating time during which heat and power turbines worked under a heat load has generally been low, though it has improved over time, as shown in the following tabulation.*

1955	34 percent
1958	39
1960	32.1
1965	41
1970	54
1971	55
1972	57.5
1973	58.7
1975	61.7

Another indication of dependence on condensing regimes is that average annual hours for utilization of heat capacity are about a thousand hours less than for electrical capacity (*Elektricheskie stantsii*, 1977:8, p. 17).

Some of the explanation for poor utilization of heating capacity is variation in heat demand, but this is far from the whole story as indicated by the fact that even today *peak* heat loads are far below heating capacities. In 1975, the aggregate heat *capacity* of all TETs was 272,000 gigacalories per hour, but the aggregate of peak *loads* was only about 200,000 gigacalories per hour. The ratio of capacity to load was

*Based on *Elektricheskie stantsii*, 1974:11, p. 6; Vilenskii, 1963, pp. 83–84; Levental' and Melent'ev, 1961, p. 17; Gorshkov (ed.), 1967, p. 51; *Elektricheskie stantsii*, 1977:1, p. 82.

higher for Minenergo (1.42) than for other stations (1.25), a difference
probably explainable by the problem of coordinating capacity and
load when heat is supplied on a utility basis (*Elektricheskie stantsii*,
1977:8, p. 16). One especially bad example cited is a TET with a
heat capacity of 848 gigacalories per hour for which the maximum
load was only 289 gigacalories; another has no heat network at all
connected with it, nor is it likely to have one soon (*Energetik*, 1976:4,
p. 37).

A second type of heat and power turbine operates under back pressure from the heat load and must shut down or operate at reduced
electrical output when the heat load is absent or reduced below the
design figure. These units, too, have had low utilization. Since the
heat load curve does not correspond with the power load curve, their
electrical capacity is thus often not available to deal with peak needs.

It now appears that a serious mistake made in designing the stations
was to provide them with extraction turbines adequate to meet peak
heat loads. Even in cases where a station's heat capacity has been
matched to its peak heat loads, it will work most of the time at a heat
load well below its maximum capacity, and must send a lot of steam
to the condensers. If such stations had been equipped with additional
boilers and hot water heaters for meeting peak heat needs, the heat extraction combines could work a much larger fraction of the time at
their design heat loads and thus save more in fuel. This weakness has
been recognized for a long time (see *Elektricheskie stantsii*, 1958:2,
p. 14), but little has been done to remedy it (*Elektricheskie stantsii*,
1977:8, p. 16). It is not clear to me whether these criticisms and the
recommendations that such peaking equipment be added are really
optimizing or whether they are only an effort to perform better by the
wrong criterion—i.e., reduction of heat rates for power.

It is interesting to ask why the heat loads are so much below those
projected by the designers. The main explanation offered is that the
information regarding industrial and housing heat demands on which
station projects were based have been inflated. When the customer
enterprises were finished and in operation, their demands were behind
schedule and were below the original requests (*Energetik*, 1976:4, p.
37). Such inflation of requests, as a kind of insurance under a physical
rationing economy, is prevalent for current inputs, and it is not surprising that it should also emerge in this longer-horizon setting.

Apparently it also happens that customer enterprises whose needs
were included in making demand forecasts have a preference for, and

end up with, their own heat plants. The minister of the electric industry, P. S. Neporozhnyi, claims that there is a further anomaly in that the bigger the industrial user, the more likely he is to be able to get permission to equip his plant with its own boiler plant. Minenergo is thus deprived of the opportunity to build larger TETs and use the larger units that would then be more economical (Neporozhnyi, 1972, pp. 161–162). He suggests that permission to build such in-plant boiler installations should require permission from Minenergo, but one can certainly understand why the Ministry of the Chemical Industry, say, would not want to put itself in Minenergo's hands on so crucial a matter. There seems to be another structural defect in that heating networks are planned, financed, and constructed under separate procedures, so that even when a TET and its customers are ready, there may be no heating network to connect them (*Energetik*, 1976:2, p. 38). Pipe is typically a "deficit" item in Soviet investment projects, and this is an important reason for delays in completing heating networks (Neporozhnyi, 1972, p. 160).

From following the discussion of policy on TETs over the years, I have the impression that until the late sixties the policy of *teplofikatsiia* was implemented without a great deal of careful optimizing. M. A. Vilenskii says that the question of the effectiveness of TETs is controversial and such echoes as we hear from these arguments seem quite unsophisticated. P. S. Neporozhnyi, the Minister, once expressed the opinion that, with growth in the efficiency of condensing stations, TETs had lost their attractiveness. Subsequently, however, Neporozhnyi seems to have been converted and is said to have taken the position that electricity should replace hot water for space heating and that process steam for industry should be produced in electric boilers! Both positions sound quite eccentric. In his explication of plans for 1976–1980, Neporozhnyi says a big effort will be made to develop *teplofikatsiia*, including atomic TETs; but he then goes on to recommend the use of atomic reactors as pure heating plants. It seems an easily grasped idea that, in any situation where heat demand is concentrated enough to justify heating from an atomic heating station, it would be still more economical to combine the heating function with power generation. Neporozhnyi came to his position as Minister via the construction side of the business rather than from a background as an electrical engineer, and the presence of a technological amateur at the top may create conditions for strategic technological biases.

Beginning with the policymaking for the 7th Five Year Plan, the *teplofikatsiia* decision seems to have been approached with a great deal more subtlety. Zolotar'ev and Shteingauz make the sensible points that the savings from combined production depend on how cheap fuel is, and that the decline in the cost of fuel with the rising share of oil and gas reduces the effectiveness of *teplofikatsiia*. They link *teplofikatsiia* to the issue of hydrostations, saying that the cheapening of fuel has made both less attractive and that designers must figure out ways to save on capital costs to justify combines (Zolotar'ev and Shteingauz, 1960, p. 139). The same kind of analysis is well developed in Levental' and Melent'ev (1961); they lay out the kind of considerations that should be taken into account in deciding on the role of TETs in the system. In particular, they mention as mistakes that must be corrected: delayed construction of the heat network, inadequate variety in the types of turbines and boilers for such stations, poor design of some of the extraction turbines, the fact that many TETs are too small and could be replaced with large ones, and the doctrine that condensing-type equipment was preferable to the back-pressure type.

By the time of the discussions for the 9th and 10 Five Year Plans, as reflected in the Pavlenko-Nekrasov and Nekrasov-Pervukhin books, from which I have cited so often, the policymakers seem to have a very clear idea of all the considerations that should guide the design of the equipment and plants and their interrelationship with systems. One example is the analysis in one of these sources of the choice of the ratio of electric power capacity to heating capacity. The newest unit, the 250 MW extraction turbine operating at supercritical parameters, has a higher ratio of electrical power output to heat output than did earlier, smaller equipment. Using the new rather than the older units to serve a given heat load can thus be thought of as a way of adding to the electrical capacity of the network. But it is a capital-expensive way to do so, compared to conventional condensing stations. Calculations show that this is justified only in regions where fuel costs are high enough that the fuel savings from the higher efficiency of super-critical temperatures justifies the extra capital investment (Nekrasov and Pervukhin, 1977, pp. 93–94).

PEAKING EQUIPMENT

The problem of equipment for meeting peak electrical loads is an instructive example of a failure to do the R and D needed to create the equipment for a recognized need. It is my hypothesis that the explanation in this case is a kind of "criterion bias" in the thinking of electric power R and D authorities or, what comes to approximately the same thing, an excessively narrowly focussed priority system in R and D operations. My hypothesis is that the R and D people have long had an excessive interest in reducing the heat rate, and when they came up against the peaking problem, where economy requires accepting a higher fuel rate in order to save on capital investment, they could not give the task the attention it merited in an overall system-optimizing context. Several Soviet sources say this more or less explicitly—one example is in *Teplotekhnika* (1971:3).

For a long time Soviet power industry planners have recognized a need for equipment to handle peak and "semipeak" loads (the latter are called cycling or intermediate loads in the United States). The diurnal fluctuation in load on a power station or system looks something like that shown in Figure 3-2a. There is an economic advantage in having in the system some units that operate continuously, fully loaded, and others to operate for perhaps only half the day, handling the increment in load that arises during daytime hours. Still more specialized equipment is needed each day for only a few hours at the time of the very highest demand. The rationale for specialized equipment comes from two factors.

First, the base-load units may not be very flexible in their load carrying capacity—a large block may not be able to shed any significant part of its load and still maintain the various combustion and heat transfer processes its operation involves.* The lower limit for 160–200 MW coal-fired blocks is given in one source as about 65–70 percent of capacity (ENIN, 1968, p. 41). Moreover, most Soviet base-load units require very long periods for starting up and shutting down. It is said that the 300 MW units (which, as will be remembered, constitute about a fourth of Soviet fossil-fired capacity) require eight hours for start-up and approximately the same for shutdown and that they cannot be operated at less than about 80 percent of their capacity when fired with solid fuel. Obviously, it is not possible to start up and shut down such units each day so that they can operate for only 8–10 hours within the 24 hour period.

The second advantage of having special equipment for peaking operations is that its design can be optimized for the smaller number of hours it will operate each year. It is possible to greatly reduce capi-

---

*And in any case low loads on this equipment raise the unit cost of power because of fixed fuel costs in operating it.

FIGURE 3-2. Measures of Daily Load Variation in Soviet Power Systems

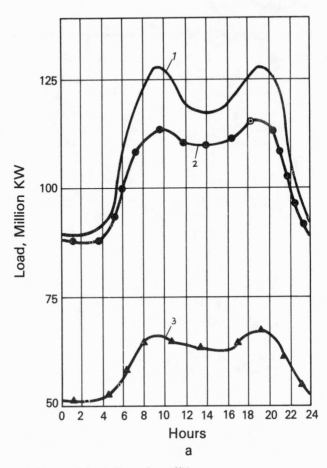

a. European grid, a weekday in December, 1975
1. Hydrostations
2. TETs and old condensing equipment
3. Nuclear and integrated blocks
SOURCE: Nekrasov and Pervukhin, 1977, p. 171.

tal cost per KW of capacity by accepting lower thermal efficiency, and, given the relatively small number of hours peaking equipment operates, this penalty may be more than made up by the saving in the opportunity cost of investment tied up in generating plant.

Though the Soviet Union for a long time had a relatively flat load curve compared to most other countries, and hence a relatively high number of hours of utilization of capacity, the situation has now changed a great deal. One measure is given in Figure 3-2b, showing the

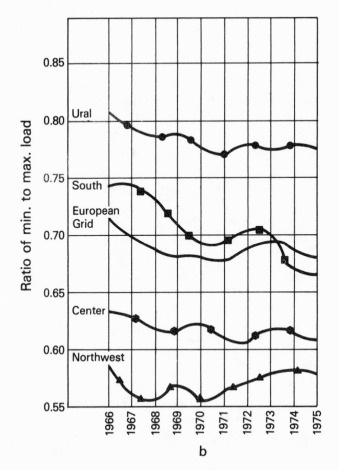

b. Change in Ratio of minimum to maximum load, 1966–1975

steady fall in the ratio of the minimum load to the maximum load for
the major power networks of the country. (I believe the Siberian and
Far Eastern networks may be an exception and so are omitted.) As
another indication, the average hours of utilization for all Minenergo
stations was 5,944 hours in 1950, but had dropped to 5,257 in 1975.

As another gross indication of the need for more flexibility in peaking capacity, Soviet power networks have very high ratios of peak load
to capacity. In 1970, the ratio was 0.81 for the European system as a
whole and within that an incredible 0.98 for the Center system. Analogous figures for the U.S. are more like 0.77–0.80 (Pavlenko and
Nekrasov, 1972, p. 198; and FPC, *Electric Power Statistics*).

On the demand side, the difference between the daily peak and low has increased in recent years because of a shift in the composition of sectoral demand, and a shortened work week has increased variation during the week. On the supply side, the declining share of hydroelectric capacity in total capacity (used in the past to meet daily peaking needs) has hurt. As one offset, the growth of regional systems and interties between them has made it possible to meet peak needs in part by transfers. One disadvantage of that solution, however, has been that in the geography of the Soviet energy sector the direction of power transfers has run counter to interregional fuel flows! (*Teplotekhnika*, 1971:3, p. 10.) The peaking problem is especially great in the European USSR where water shortages place an additional constraint on the hydro answer, and where there are few big continuous users like aluminum plants to raise the share of base-load demand. Such users have been located in Siberia to take advantage of cheap labor (Styrikovich, in *Energetika i transport*, 1973:2, p. 6). The problem will become even more serious as the share of nuclear power rises, especially in the European part of the USSR. These nuclear plants are intended for base-load operation. Technically, they can handle some load variation; but one type of reactor has been chosen specifically to accumulate plutonium to stock the first generation of breeder reactors, and variable operation greatly reduces its plutonium-producing potential.

The lack of adequate capacity to meet peak loads leads to cutting off customers at high demand periods, and voltage drops below the established standards. The problem has been discussed for many years, but the task of developing peaking equipment has been put off. In typical Soviet fashion, such development work as was done on peaking equipment was concentrated for a long time on a single solution—gas turbine equipment. Some development work was done on pumped storage, but Soviet power officials were pessimistic about the economic advantage of that approach compared to gas turbines. Only recently has any attention been given to design modifications of large boilers and steam turbine units that would make them easier to start up and shut down. Another possible solution not yet far advanced is combined gas-steam turbine cycles. The history of each of these efforts is revealing about the Soviet innovation system; let us consider them in turn.

Gas Turbines

The development of gas turbines has been slow, with many technical setbacks. Work started in the USSR to create this technology in the late fifties, at about the same time as in the United States. But whereas the United States quickly mastered and diffused gas turbine facilities for peaking, the USSR has not been successful in creating this technology.

The program for gas turbines, when it was originally set up, envisaged development of the following series of units (Sivakov, 1968) :

UNIT	PRODUCER
GT–12–3–650	Nevskii MZ (Neva Machinery Plant)
GT–25–700	LMZ (Leningrad Metal Plant)
GT–35–770	Khar'kov Turbogenerator Plant
GT–60–750	LMZ
GT–100–750	LMZ
GT–200–750	LMZ

Gas turbines were first used for electric power generation in connection with underground gasification of coal in the Moscow basin, using the 12 MW unit. One of these was installed in the Shatskaia station near Tula and brought to full power in 1958 (*Energomashinostroenie Leningrada v 1959–1965 gg*, p. 14). Two such units were built, but "this equipment received no further development," because of the problem of cleaning the gas, which turned out to be too expensive (*Energeticheskoe mashinostroenie, 1917–1967*, 1967, p. 55). The 12 MW model was also used in a station at Nebit-Dag, where four units were put into operation in 1964–1968 (*Teploenergetika*, 1970:11, p. 31; *Energeticheskoe mashinostroenie, 1917–1967*, 1967, p. 56; *Energetik*, 1975:9, pp. 16–17). Not much has been said about these 12 MW turbines in later discussion, but it is clear they were not designed as peaking units at all; they just represented electric power applications of the units that the Nevskii machine-building plant was producing for compressor stations. Their justification at Nebit-Dag was a plentiful supply of oil-well gas; another advantage in that desert environment was that their requirements for cooling water were small.

The literature on the 7th Five Year Plan (1959–1965) says that two larger experimental units—one 25 MW and one 50 MW—are being developed (Abramov, 1959, pp. 35–36). LMZ was to produce the 25 MW unit of the original plan and the Khar'kov turbogenerator plant, a 50 MW unit (Zhimerin, 1960, p. 181), probably as a modification of the originally assigned 35 MW unit. These were intended for power plants in Kiev and Khar'kov respectively with experimental operation planned to begin in 1962 (Novikov, 1962, p. 231). Both these projects encountered serious troubles. Both units were supposedly installed in 1963 (*Teploenergetika*, 1970:11, p. 3), but as late as 1968 a special decree of the Council of Ministers regarding the electric power industry complained that both were behind schedule and still not ready for operation. (The decree is given in *Resheniia Partii i pravitel'stva po khoziaistvennym voprosam*, vol. 6, Moscow, 1968, pp. 643–655.) By 1970 the 25 MW version had operated for 14,000 hours, and the 50 MW unit was in experimental operation (*Teploenergetika*, 1970:11, p. 3). Apparently the 50 MW unit has not been a success, and nothing more is being said about it, though the Khar'kov plant is still working on gas turbines and has produced a 35 MW unit for use in a steam-gas combined-cycle unit (Pavlenko and Nekrasov, 1972, p. 239).

The 25 MW size was sufficiently successful to encourage further use. LMZ was reported to be producing several 25 MW units to be used in the Iakutsk GRES (Neporozhnyi, 1970). The first of four units was started up in 1970, and a second was installed and tested soon after (*Energomashinostroenie*, 1970:4, p. 5). But the 25 MW units are probably best considered as an experiment. It is quite possible that they are not expected to play a peaking role but are being used to take advantage of low cost gas for a small station. LMZ is also supposed to be producing a 30 MW unit that can be used either as peaking or base-load equipment, which sounds like a dubious notion, since design needs to be optimized according to quite different criteria for the two purposes.

The next step and the main focus of current effort is the 100 MW unit assigned to LMZ for production and intended for use in the Krasnodar GRES. In the 1968 decree mentioned above, the Council of Ministers directed that this 100 MW unit be finished and started up no later than 1970. A 1969 source reports it as already produced and in the process of being installed (*Vestnik mashinostroeniia*, 1969: 12, p. 3). But there have clearly been great difficulties with getting it into operation, and in 1975 it was reported as still in the process of being mastered (*Energetik*, 1975:2, p. 5). The 9th Five Year Plan envisaged that six such units would be commissioned (Baibakov, 1972, p. 101), but so far as I can tell no additional units were in fact commissioned in the 9th Five Year Plan. It is said that LMZ has significantly improved the 100 MW turbine on the basis of the experience with the Krasnodar plant, and is to produce ten such turbines in 1976–1980 (Nekrasov and Pervukhin, 1977, p. 201).

For completeness I should add that there is a miscellany of smaller gas turbine units in use in electric power stations. On 1 January 1968, there were 30 gas turbine units in operation for electric power generation with a capacity of 300 MW (P. S. Neporozhnyi, *Elektrifikatsiia SSSR*, Moscow, 1970, pp. 2, 5). That means an average of 10 MW each, and most seem to be intended for standby emergency use. There is also a ship-base generating station with a total capacity of 22 MW using gas turbines created by the shipbuilding industry (*Ekonomicheskaia Gazeta*, 1976:2). Here the rationale is mobility rather than peaking use.

In addition to the slow progress in developing and mastering the experimental gas turbine units, it seems that the models so far produced are not really suitable for the intended purpose. The 100 MW unit was the first gas turbine unit supposedly designed for peaking use (*Energeticheskoe mashinostroenie, 1917–1967*, Moscow, 1967, p. 56). According to Academician M. A. Styrikovich, however, even this unit is not really suitable for peaking, since it is quite elaborate and expensive in design (to achieve fuel economy) and has a high capital cost per KW of capacity. Moreover, it requires 35 minutes from a cold start to full load, which Styrikovich contrasts with a U.S. 150 MW unit that can do so in half the time (*Energetika i transport*, 1973:2, p. 7). The Styrikovich claim may be overoptimistic—two of these units installed in a power plant in Hungary require 40–45 minutes to be brought to full load (*Energetik*, 1976:6, p. 29).

The Hungarians report a heat rate of 500–514 grams of standard fuel per KWH, and the Krasnodar unit is said to have experienced a rate of 500 grams/KWH (Karol', 1975, p. 30). Since the U.S. rate for gas turbine plants is about 545 grams/KWH (FPC, *Gas Turbine Electric Plant Construction Costs and Annual Production Expenses*), one wonders whether the designers may have gone too far in trying to save fuel, though it could be that they are only responding to a higher relative cost for fuel. A failure to optimize for peaking purposes would be understandable since LMZ has specialized in producing turbines for compressor stations, which involve continuous use. Also, all these gas turbine units have been combined with standard generators, rather than with special generators optimized for the low number of hours of utilization (Sivakov, 1968, p. 39).

The contrast of this history with U.S. experience is striking. The first U.S. gas turbine plant with a capacity of 300 MW was installed in 1958. The FPC first began reporting on gas turbine units in U.S. utility plants in 1963, when there was a total of 600 MW of installed capacity, and by the end of 1975 total installed capacity was 43,533 MW. And by that date this included single *units* of 300 MW capacity.

If the plan for producing and installing 10 LMZ–100–750 sets in the 10th Five Year Plan is actually achieved, Soviet capacity at the end of 1980 would not exceed about 1,500 MW. In view of the long delays compared to targets for individual programs (five to ten years) and the long lag behind the diffusion of this technology in the United States, my conclusion is that the Soviet gas turbine program has thus far been a failure.

I suggest several factors to explain these failures in the gas turbine program: the whole Minenergo structure has not put a high priority on peaking problems; the plants and designers producing turbines are much more concerned with turbines for compressor stations and other continuous uses than with turbines for electric power generation; there seems to be a specific technological problem with metallurgy and with reliability that is also evident in the gas turbine program for compressor stations.

Pumped Water Storage

An alternative for meeting daily peaks is pumped storage, in which a system's unused capacity in off hours is used to lift water into an artificial reservoir from which it will flow back to generate power in the peak periods. Investment costs for construction and equipment can be reduced by using a reversible unit as a motor-pump combination in one part of the cycle and as a turbine-generator unit in the other. The research organization Gidroproekt imeni Zhuka began studies on pumped storage stations in 1959, and work was begun in 1963 on the first station, near Kiev, which came into operation in the early seventies (ENIN, 1963, p. 257; and Karol', 1975, p. 3). One of the main functions of this project was to test the new reversible units for this kind of work. The station has a 200 MW capacity in six units, of which three are reversible, with pumping power of 42.4 MW each (Gidroproekt, 1972, p. 182). How successfully this station has operated is unclear—a recent survey of the Ukrainian power industry mentions it but gives no details (Shvets, 1970, p. 112).

Soviet analyses of the costs of pumped storage for a long time concluded that pumped storage was not competitive as a peaking approach. One source asserts that its capital cost per KW is higher than for other types of peaking units and is "approximately on the same level as base-load condensing stations" (Sivakov, 1972). The approach used is to figure a permissible cost for pumped storage equipment that would not raise costs in the system above that which would result from using standard condensing steam stations instead (i.e., a kind of shadow price); Sivakov says that the projected cost of the pumped storage capacity would be somewhat above this. Nevertheless the need is urgent, and a useful and more recent comprehensive summary of Soviet analyses of pumped storage (Karol', 1975) suggests a more favorable evaluation of its competitive position. The 10th Five Year Plan guidelines direct Minenergo to accelerate the construction of pumped storage stations (*Ekonomicheskaia Gazeta*, 1975:51, p. 6). A 1,200 MW station is now being built on the Kunia River near Zagorsk (*Energetik*, 1974:4). For future pumped storage it is apparently intended to use more powerful equipment—LMZ is to produce the *golovnoi* 200 MW reversible unit during 1976–1980 for the Zagorsk facility (Nekrasov and Pervukhin, 1977, pp. 200–201).

It is very important to find some solution for this peaking problem, especially in the European USSR where a large number of nuclear stations are being built and where there is little unused hydroelectric potential. One source indicates that pumped storage is expected to be used primarily to complement base-load nuclear plants (*Teplotekhnika*, 1971:3, p. 5).

"Semipeak" Equipment

In addition to sharp morning and evening peaks in the daily demand pattern there is a daytime/nighttime difference that calls for equipment to be operated for half the day or more, but to shut down for the other half. As the Russians have shifted to 300, 500, and 800 MW block units, this need has been ignored and the large blocks must be kept in operation even at low loads just because they are too difficult to start up and shut down within the daily cycle. The research organizations originally recommended that a simplified medium-size unit with steam at 160 atmospheres and 520–540°C would be the best bet (*Teploenergetika*, 1971:3, p. 10); but the actual decision has been for a 500 MW unit with steam at 130 atmospheres and at 510°C for both initial and reheat stages, burning either gas or residual fuel oil. It would be started and shut down about 300 times per year and would work about 3,000–3,500 hours (*Energetik*, 1974:8, p. 4—the detailed rationalization of the specifications is explained in an article in *Teploenergetika*, 1975:5, pp. 11–16). The new unit is calculated to have a capital cost of 95 rubles/KW and a fuel expenditure of 370–375 grams/KWH. The guidelines for the 10th Five Year Plan reemphasize the importance of such equipment, directing that the electric power industry "accelerate the mastering of highly flexible 500 MW generating blocks" (*Ekonomicheskaia Gazeta,* 1975:51, p. 6).

LMZ has been working on the prototype since the end of the 9th Five Year Plan, and it is to be produced in the 10th (Nekrasov and Pervukhin, 1977, pp. 198–199). No clues as to current progress have been found, but it is not likely that this new equipment will be ready soon. According to one author: "work on equipment for semipeaking loads is in an embryonic stage. . . . The work on creating equipment adapted to work in the semipeak part of the load curve must be forced, since the need for such capacity is already 5 million KW" (*Energetik,* 1975:2, p. 5).

Combined Cycles

Finally, it is hoped to combine gas and steam turbines in equipment which will both have peaking potential and offer big fuel economies.

The R and D effort was focussed at first on a cycle in which gas was burned under high pressure in a steam generator, with the combustion products passed through a gas turbine and feed water heating equipment while the steam was used to run a traditional steam turbine. The first such unit of 16,000 KW capacity was commissioned in 1965 at a Leningrad station; it consisted of a 4,000 KW gas turbine and a 12,000 KW steam turbine (*Leningradskaia promyshlennost' za 50 let,* 1967, p. 362). It took a long time to make this unit work. As late as the decree of January 1968 on the power industry, the Central Committee and the Council of Ministers complain that this experiment was not on schedule. A couple of other experimental units were also installed in the sixties. On the basis of these experiments it was decided to produce a 200 MW unit that would include a 30–40 MW gas turbine, to be installed in the Nevinnomysskaia power station in 1971. There was some slippage, but tests with natural gas began in 1972. This unit was intended for use with either fuel oil or gas and tests began in 1974 to develop the fuel oil modification (*Teploenergetika,* 1975:6, pp. 27–28). There were reliability problems, but the equipment operated for a significant amount of time and was considered a success. The tests revealed a fuel rate only about equal to supercritical steam units, but the combined-cycle equipment saves on capital costs and could be used for semipeaking operations.

Apparently the original intention of using gas in these units was not really practical in Soviet circumstances in most areas, because gas could not be assured reliably year round, or would require big investment costs for delivery. Hence, a decision was made to design the units to operate on fuel oil as well. Ironically, about the time the tests on the Nevinnomysskaia prototype were completed (mid-seventies), it became clear that oil was a deficit fuel, and it became a policy goal to avoid use of fuel oil in new power plants. Hence, it is not intended to make extensive use of these units in the 10th Five Year Plan. Some possible application is seen in Tiumen' *oblast'* using surplus oil well gas and in Central Asia on the basis of local gas fields. For these applications, what will apparently be a new round of development work will go ahead on a 250 MW version (Nekrasov and Pervukhin, 1977, pp. 108–109).

The combined-cycle idea can be embodied in an alternative form that can burn mostly coal. In this version, gas or fuel oil is burned to run a gas turbine and the gases are exhausted from the turbine to a traditional boiler burning coal. These gases contain both heat and oxygen from the excess air introduced to maintain the gas temperature at the turbine inlet at no more than 750°. This form is desirable in the new circumstances, since it offers the prospect of getting a unit for semipeaking operations at no fuel penalty compared to the large condensing units. This direction, too, will require starting a development program anew.

With the benefit of hindsight, one is tempted to say that the combined-cycle program was bungled in the sense that the R and D effort was pursued on the basis of a very mistaken forecast about objectives. On the other hand, it may be too much to expect that Soviet R and D planners should have seen in the mid-sixties the shift in the fuel balance that would occur ten years later, thus saving themselves the trouble of creating a technology that could not economically be used. Still, this case clearly shows once again that the long range vision sometimes hypothesized as guiding Soviet fuel policy and energy R and D policy (and offered as a contrast to our own) may also stumble over mistaken forecasts.

CONCLUSIONS

To conclude this chapter, it may be helpful to draw together some intermediate level generalizations and questions about Soviet R and D practices revealed in the cases we have described. They all involve the creation of a very elaborate item of equipment—the 300 MW generating unit, for example, includes a boiler, turbine, and generator as main items plus a great deal of auxiliary equipment—which must in addition be fitted into a larger system. Moreover, major elements will be produced in different plants—the boiler in one organization, the turbine in another, and the generator in yet another. There is thus an extraordinary integration task to be performed, both at the original design stage and at the test and adaptation stage. I have an impression that in approaching this problem, the Soviet system goes rather directly to the creation of what is called a *golovnoi obrazets* at a fairly early stage and relies on this to settle many issues that might have been settled earlier if more effective design work were done. This *golovnoi obrazets* might be called a prototype, though it seems to be preceded and accompanied by too little development work even to assure that it will work when installed. It must be supplemented with makeshift auxiliary equipment; significant components of that prototype (such as the boiler or a whole stage of the turbine) may be replaced in an effort to make it work once installed. The history of all these examples shows a long period of *naladka*—i.e., adjusting, adapting, fixing, so that the item will function. The same pattern will be observed in cases of nuclear power R and D and magnetohydrodynamic power generation. One would think a great part of these delays and this expense could be avoided by more careful testing of components, more careful scheduling of the whole process in advance.

But this *golovnoi obrazets* is not a prototype in the sense of being just a test or test-bed—it is a full fledged, expensive piece of equipment intended to be used commercially. This is what happens in any country to some extent, but I suspect that U.S. firms would do a great deal more preliminary work and would then expect the new piece of equipment to work essentially as the designers intended.

Another aspect of the Soviet approach is that a new development effort is often limited to a part of the system, relying on standard associated equipment or components to round out the system. Thus, the peaking units supposedly now being developed will use standard generators rather than new designs optimized for peaking. The 250 MW heat and power unit was consciously designed to make it possible to use as auxiliary equipment that already developed in connection with the 300 MW condensing blocks.

The telescoping of successive phases seems to continue through later stages, as well. For example, the Soviets started producing the 300 MW block in series before it had been perfected to the point where it could be delivered, installed, integrated, and made to work with the kind of predictability one would expect for a "developed" technology, embodied in equipment produced in series. Sometimes the problems never get ironed out. I assume that the notion that all the models from the 150 MW units on would operate at a temperature of 565° must involve significant wastes when the original designs continued to be produced, only to be universally operated at lower temperatures.

It is interesting to think about the costs and benefits of this strategy. It is my hypothesis that the USSR could produce a usable, acceptable, model more quickly, have the models they produce in series work better, and so on, if they did more testing and more careful design work first. On the other hand, it may be that plunging ahead will be an effective way to identify problems, focus effort on crucial issues, and accelerate the transition to commercialization. One way to think about it is to say that this rapid push to commercialize is advantageous in the sense that it shows what the problems are and reveals what adaptations need to be made. But the feedback cycle may be too slow to take advantage of this supposed virtue. Thus, the Soviets consistently found that because the coal delivered to power plants had less than the expected heat content, the boiler outputs were often insufficient to fully utilize the capacity of the 300 MW turbogenerator units. So far as I can tell, however, this new intelligence had no effect on the design of the equipment; at least the impact is delayed for a long time. It is interesting to find one Soviet commentator expressing the idea that there are big costs to the Soviet way and that it is wasteful:

> The technical base of experimental facilities often does not permit the proper development of designs for machines. . . . Thus, because of the lack of test equipment for developing the *golovnoi obrazets* of the 300 MW turbine, at the Khar'kov turbine plant and the generator at the Elektrotiazhmash plant, the losses to the national economy exceeded by several times the cost of construction and equipping test facilities. [ANSSSR, 1969, p. 166]

In other words, the Soviet approach is perhaps less a well-thought out R and D philosophy than merely an expedient, faute de mieux.

The power industry people are aware of the problem and are trying to change it as fast as they can. Nekrasov and Pervukhin describe big expansions of test facilities at most of these plants, and the Soviet economist Efimov, noting the adverse effect that a poor experimental base has on Soviet R and D generally, notes Elektrosila and the Khar'-

kov turbine plant as exceptions where adequate facilities ensure testing along the way in the test process with desirable results (*Planovoe
khoziaistvo*, 1974:11, p. 18).

But I think the problem has additional dimensions. I am not sure
that the developers see experimental operation and debugging as a
way to help them improve, perfect, and optimize the technology. They
may think they have done their job when they have delivered the
golovnoi obrazets and may resent rather than welcome any feedback
that tells them what changes must be made. They may not even be
much involved in the process. In its 1968 decree on power industry
problems, the Central Committee and the Council of Ministers made
a special point of saying that responsibility for mastering the *golovnye
obraztsy* of new equipment is not the task of Minenergo alone. Some
responsibility must also be assumed by the ministries, plants, and design organizations that produced the equipment (p. 648). I feel that
one of the reasons Minenergo has done relatively so well among the
various energy sectors in achieving technological progress is that it has
the organizational structure and the bureaucratic power to act as its
own general contractor, not just for investment, but for the associated
development processes as well. This has enabled it to achieve a toler-
able level of integration. I suspect that one of the things crucial to its
success is that it can get most of the elements of the core technology—
the turbines and the generators—out of a relatively small number of
very large plants.

4

Coal Mining

THE TECHNOLOGICAL LEVEL OF
SOVIET COAL MINING

As Chapter 2 indicated, the Soviet R and D effort in coal mining seems to be very large and long established. Large amounts of resources have been devoted over the years to improving the technology of this industry, suggesting that it ought to be reasonably advanced. In fact, however, coal is the branch of the energy sector in which the USSR is comparatively most backward. Table 4-1 shows some comparative data on the output and inputs of the U.S. and Soviet coal-mining industries as a basis for inferences about comparative productivity and technical level. The comparison is made for 1972, since that is the most recent year for which data from the U.S. Census of Mineral Industries is available; a comparison for 1975 would be slightly more favorable to the USSR. This comparison is offered only as a starting point and will have to be disaggregated to be very instructive. The U.S. and Soviet coal industries are quite different in the relative importance of underground and open-pit mining and in the conditions of mining and technologies used within each of these two major divisions. This will become clear as the comparison proceeds.

In the early seventies, labor productivity in Soviet coal mining was only about 15 percent as high as in U.S. coal mining, when Soviet output is adjusted to a cleaned basis.* The Soviet Union reports employment of around a million persons in coal mining compared to 159,000 persons in the United States, who produce somewhat more coal. In fact, Soviet relative labor productivity is still lower than this comparison suggests, since the U.S. figure includes all employees, whereas the Soviet figure includes only those engaged in coal-mining operations and omits large numbers of workers in repair, transport, construction, and other auxiliary operations.

---

*The Russians sometimes report a figure for net output (i.e., after cleaning), but even that measure needs to be further adjusted downward for the fact that some raw coal is cleaned by the Ministry of Ferrous Metallurgy rather than the Ministry of the Coal Industry and that over half of all coal is shipped to customers, especially to electric power stations, without cleaning. Also, much of what is counted as clean coal (middlings, slurry, screenings) contains a large amount of rock.

TABLE 4-1. Comparative U.S.-Soviet Indicators for Coal Mining, 1972

	USSR	U.S.	U.S./USSR
Output			
Natural tons (metric) as reported (millions)	655.2	546.6	0.83
Underground mines	465.0	268.7[a]	0.58
Strip mines	190.2	253.3	1.33
Corrected to cleaned basis (million metric tons)	524.2[b]	546.6	1.04
Heat content as reported (million tons of standard fuel)	459.8	522	1.14
Employment			
Number of employees (thousands)	1,056	159.3	0.15
Number of production workers (thousands)	823.5[c]	129.3	0.16
Capital			
Fixed assets	11.6BR[d]	5.5B$	—
Buildings and structures	8.1[d]	1.0	—
Machinery and equipment	2.9[d]	3.8	—

Underground equipment

Continuous mining machines and long wall cutters	4,107	1,883	0.46
Coal cutting machines	359	1,890	5.26
Loading machines	4,518	1,959	0.43
Electric locomotives	13,000	3,412	0.26
Rubber-tired tractors	neg	1,937	—
Shuttle cars	neg	6,367	—

SOURCES AND NOTES: Except as noted, Soviet data from standard handbooks; U.S. data from either 1972 *Census of Mineral Industries* or 1972 *Minerals Yearbook.*

aIncludes auger mining.

bIn 1972, 47.3 percent of coal was cleaned and, for the part cleaned in *Minugol*, losses and rock were 15.7 percent. Another 27.5 percent was middlings, fines, and screenings, with a high rock content. Ash content of all coal shipped, including that which had been cleaned, was 20 percent, so as an approximation I deduct 20 percent of raw coal output to clean it to U.S. levels. The reduction from raw to clean coal in the United States in 1972 according to the *Minerals Yearbook* was 27 percent, so that the adjustment for the USSR is probably conservative.

cFigured by dividing average monthly output per "worker engaged in coal mining" (given in *Ugol'*, 1973:4, p. 74) into output. Soviet employment figures are probably highly restrictive in terms of activities covered and (in the case of production workers) in the categories of workers covered. In terms of activities, coal cleaning, briquet production, capital repair, construction transportation, and other such "nonindustrial" activities as exploration and marketing are excluded. As an illustration of the excluded activities, in the Kuzbass combine in 1957 total employment was 1.8 times employment in the narrow definition of coal mining. Though U.S. mines are probably likely to obtain some of these services (such as construction) from outside firms, the U.S. employment figure shown is more inclusive than the Soviet one in terms of activities, since it includes all employees of the mines. It is estimated that total employment in the coal industry today is nearer two million than the roughly one million persons reported in the handbooks (Strishkov, 1973, p. 48).

dBased on official handbook data for total fixed assets in industry and percentage breakdowns by sector.

The USSR also has a very large investment in fixed assets in coal mining compared to the capital stock of U.S. coal mines. The 1972 Census of Mineral Industries shows the value of the capital stock in U.S. coal mining as 5.5 billion dollars, while the Central Statistical Administration reports capital stock in Soviet coal mining as 11.6 billion rubles. Since most studies show that the ruble cost of one dollar's worth of fixed assets is much less than one ruble, capital productivity in the USSR would also seem to be quite low.* Some further information on the structure of these two aggregates helps explain how the difference can be so great. About 70 percent of the U.S. total consists of machinery and equipment, whereas in the Soviet total less than 30 percent represents machinery and equipment. Most of the rest in both cases consists of buildings and structures (and in the U.S. case, investment in the development of mineral properties). The very heavy investment in buildings and structures reflects partly the higher relative share of underground mining in the USSR (70 percent compared to 54 percent in the United States in the mid-seventies) and the deeper, more complicated, conditions of Soviet mines. It may also be that development work is more often capitalized in the USSR than in the United States, where it is often expensed. (According to the 1972 *Census of Mineral Industries*, 26 percent of outlays on mineral development and exploration made by the coal industry in 1972 were expensed, rather than capitalized.) But even if we look at machinery alone, the fact that most studies show a ruble/dollar ratio for machinery prices well below 1 (a ruble may buy twice as much machinery as a dollar) implies that the Soviet Union uses a larger stock of machinery than the United States to produce a smaller output. Examining that conjecture by looking at actual numbers of various kinds of machines reveals more about the differences in mining techniques than about relative holdings of machinery stocks.

Given the difference in the relative importance of strip and underground mining in the two countries and its impact on overall productivity, a useful comparison requires that we study strip and underground mining separately.

Comparison with U.S. Strip Mining

There are some differences in the conditions even of strip mining of

*CIA, 1976; Treml and Gallik, 1973. The latter source suggests a ruble/dollar ratio for machinery in the range 0.5–0.7.

coal between the United States and the USSR. The Soviet Union has relatively few hillside open pits worked on the contour and relatively few flat-lying seams like those in the U.S. Midwest. But there is a significant core of technology common to both countries that is more or less independent of natural conditions; hence labor and equipment productivity comparisons reliably support some inferences about performance on technology and innovation. The major operation in open-pit mining is excavation—removal of overburden and coal—mostly by various kinds of excavators. Since the comparative volume moved in the two countries is very close to their relative standing in aggregate capacity of single-bucket excavators (see table 4-2), the productivity of excavators in the USSR appears to be about as high as in the United States. This is only an approximate comparison, since the United States supplements its excavators with considerable numbers of front-end loaders, scrapers, and bulldozers, all of which the USSR has in much smaller numbers. On the other hand, the USSR removes a lot of its coal with wheel excavators (23 percent of all coal produced was removed with such excavators in 1973) and puts part of the job of overburden removal on trains, which are virtually never used in the United States. Comparatively good capital productivity is also suggested by the fact that it would take a ruble/dollar ratio of 0.33 for capital to make capital stocks in the two countries proportional to the amount of coal and overburden removed. (I indicated earlier that the actual ratio might be something like 0.5–0.7.) Labor productivity, however, seems quite low, at 25 percent of the U.S. level, even though better than in all coal mining. However, when we take account of the much larger amount of overburden removed in relation to coal in the United States and figure labor productivity in terms of total volume removed, the Soviet level falls to 16 percent of the U.S. level.

Underground Mining

In underground mining there are very large differences between Soviet and U.S. mining conditions and technologies employed, and it is more to the point to make comparisons with Western European mining.

Most U.S. underground coal mining uses the "room and pillar" method, in which coal is removed from relatively short faces either by the "conventional" technique (coal is removed from the face by undercutting and blasting and is then loaded onto shuttle cars) or by con-

TABLE 4-2. Comparative Indicators for U.S. and Soviet Open-pit Coal Mining, 1972

	USSR	U.S.	U.S./USSR
Output			
Natural tons (metric) as reported (millions)	190.2	253.3	1.33
Natural tons corrected to cleaned basis (millions)	152.2[a]	253.3	1.66
Overburn removed (million cubic meters)[c]	727.5[b]	2,130.3[a]	2.93
Total volume removed (million cubic meters) [d]	938.8	2,475	2.64
Employment			
Number of employees (thousands)	61.3[e]	25.9[f]	0.42
Number of production workers (thousands)	47.8[e]	22.1	0.46
Capital			
Fixed Assets	1.23BR[g]	1.1B$	—
Equipment			
Number of single-bucket excavators	1,356	3,416	2.52
Aggregate bucket capacity (cubic meters) [h]	7,187	18,788	2.61
Bucket wheel excavators	35[i]	9	0.26
Bulldozers	741[j]	3,891	5.25
Carryall scrapers	neg	360	—
Front-end loaders	neg	2,211	—
Locomotives	812[k]	neg	—

SOURCES AND NOTES: Except as noted, Soviet data are from standard handbooks, and U.S. data are from the 1972 *Census of Mineral Industries* and the 1972 *Minerals Yearbook*.

aData on losses in cleaning are not available separately for underground and strip mining, so I have used the 20 percent correction explained in the notes to Table 4-1.

bN. V. Mel'nikov in *Ugol'*, 1976:2.

cThe U.S. overburden/coal ratio in 1965 was 12.8 cubic yards per short ton, and was forecast to be 10 years per ton in 1985 (U.S. Bureau of Mines, 1972, p. 6; and U.S. Bureau of Mines, 1976, pp. 27–28). I have taken 11 yards per ton as a plausible interpolation for 1972.

dRaw coal plus overburden, counting coal at 0.9 tons per cubic meter, which seems a representative figure in the detailed data given in Ulianov, 1972. In the United States, clean coal output has been adjusted to a raw coal basis by the ratio indicated in the notes to Table 4-1.

eAn estimate of production workers can be based on the output and average monthly productivity figures given for all coal mining, underground mining, and strip mining (respectively, 66.3, 50.5, and 335.1 tons per man per month) in *Ugol'*, 1973:4, p. 74. I distribute total employment given in Table 4-1 in the same proportions.

fThe *Census of Mineral Industries* does not break down all employees between strip and other mining, so I use the same ratio of all employees to production workers as for all coal mining.

gBased on *Ugol'*, 1974:11, p. 48.

hAverage bucket size for Soviet excavators can be figured as 5.3 cubic meters in 1970 on the basis of Zhuravlev (1971, p. 7), and I assume that it was the same in 1972. The U.S. capacity can be estimated only approximately on the basis of a class distribution in the *Minerals Yearbook*. Assuming that the 2,418 excavators in the class "less than 6 cubic yards," have an average of 3.5 cubic yards, and that the average for the 47 in the class "more than 50 cubic yards" is 70 cubic yards, and using midpoints for the other classes, the average is 5.5 cubic meters.

iBased on N. V. Mel'nikov in *Ugol'*, 1976:2.

j1970: only active units are included (Dobva, 1978, p. 12).

kZhuravlev, 1971; refers to 1970, and the 1972 figure would be somewhat larger.

tinuous miners which combine the winning and loading processes. European (and Soviet) mines generally use the "long-wall" method, in which larger panels are totally excavated by means of continuous miners moving across a long face of the panel, with the coal being removed to a haulway by a face conveyor. The long-wall method involves letting the roof collapse behind the working; the U.S. method keeps the roof intact, usually through bolting. Soviet underground mines are deep, tend to have a high proportion of relatively thin seams, are gassy, and in many cases are subject to rock and coal bursts. The average depth of seams being worked in the USSR is about 331 meters, in West Germany 756 meters, and in France 564 meters. (The usual Soviet statement that their depth conditions approximate those of France and Germany is more nearly true for the Donbass than for other regions.) The average thickness of seam being worked in the USSR is 1.35 meters (1.05 in the Ukraine), compared to 1.65 in the United States, 1.3 in Great Britain, 1.54 in West Germany, and 1.59 in France. In the Donbass, 61 percent of all faces being worked are a meter or less in thickness, and a fifth of them dip at more than 18°.

A detailed comparison of Soviet underground mining with Western European mining is too complicated a task to undertake here, but a comparison of labor productivity will give some clue to the relative technical level of Soviet underground coal mining. Care must be used in such comparisons because of differences in the way productivity is calculated. Soviet statements about labor productivity in coal mining usually refer to average monthly output per worker, with output measured as raw coal, and with workers restricted to "production workers engaged in coal mining" (*rabochie po dobyche uglia*). Most European data use clean coal as the output measure and have a much larger coverage of workers in auxiliary activities. The Russians have reported some productivity data to the Economic Commission for Europe, purportedly adjusted to standardized concepts, for underground hard coal mines (ECE, *Annual Bulletin of Coal Statistics for Europe*, various years), though experimentation with these numbers does not reassure one that they are really in terms of the ECE concepts; they may cover a more restricted range of workers and the Soviet adjustment to a cleaned basis is inadequate, as explained earlier. In 1970—the last year for which the Russians reported these data—average monthly output in tons per man in hard coal underground mines was as follows:

USSR	37.7
West Germany	57.8
United Kingdom	44.4

France	31.7
Belgium	24.2
Spain	18.7
Turkey	21.9
Poland	42.8
Czechoslovakia	38.4

The USSR ranks well below West Germany and the United Kingdom, among the important coal producers, but above France and Belgium, not to speak of Spain and Turkey. Note also that Soviet productivity is below that of Czechoslovakia and Poland. In interpreting these productivity figures it is important to note that Soviet coal miners work considerably more shifts per year than miners in the Western European mines. The average annual number of shifts per worker in 1970 was 260 in the USSR, compared to 215 in the United Kingdom, 209 in West Germany, 183 in Belgium, and 219 in France (these data are also from the ECE reports). So the relative standing on a man-hour or man-shift basis, which relates more directly to the issue of relative technological level, would be less favorable to the USSR—it would make West German labor productivity almost double Soviet productivity, for example.

Since we are concerned as much with technical progress as with technical level, it is also interesting to consider the dynamics of labor productivity. In the last two decades, productivity has risen less rapidly in the USSR than in the major Western producing countries (UK, Germany, France) so that the Soviet relative standing has deteriorated.

Implications of Productivity Differences

Even in these disaggregated productivity comparisons, Soviet labor productivity is relatively low, and there are apparently two major factors that explain this situation. These come across well in a report that B. V. Bratchenko, Minister of the Coal Industry, made on his return from a visit to U.S. coal mines (*Ugol'*, 1970:2, p. 58). His group was impressed with the equipment they saw and with the simplicity of labor organization. What struck Bratchenko about labor in American mines was that it was focused on getting the coal out and was not burdened with the highly subdivided and functionally overspecialized organization characteristic of Soviet mines. He was also struck by the high capacity of machinery (e.g., large unit powers of combines, size of shovels and draglines) and with the reliability, quality, and design of the machinery for specific situations. From the other side, an American delegation that visited Soviet mines reported that "it was our impression that the mines were overmanaged and overengineered, and yet in some respects were lacking in use of available and known technology" (*Mining Engineering*, July, 1974, p. 65). Western European observers, too, have commented on the slow adoption by the USSR of some of the more modern kinds of machinery, such as narrow-web cutters and movable hydraulic roof supports, which were introduced in the USSR years after they were in Western Europe. In short, Soviet labor productivity in coal mining is low in part because of poor organization and management and little concern with economizing on labor. But there is also a more specifically technological weakness—i.e., slowness in designing, producing, and introducing improved machinery. It should be illuminating, therefore, to look at Soviet experience in dealing with some major equipment innovations, and then to try to discover which problems have interfered with more rapid modernization of machinery.

The number of types of equipment important in determining productivity in coal mining is very great, and it would take an extensive research effort to deal with them all. The approach here is the relatively modest one of looking at a few illustrative cases as a way of getting at what seem to be general systemic features, rather than trying to be comprehensive in terms of equipment categories. We shall look at technological evolution separately for underground and strip mining.

TECHNOLOGICAL PROGRESS IN UNDERGROUND MINING

There is a great variety of technologies that might be considered in underground mining—mine construction, transport, ventilation, safety, and so on. But I shall concentrate on only one part of the picture—the coal extraction process at the face and the delivery of the coal to the haulage point.

As already mentioned, most flat and gently sloping seams in Soviet mines are worked by the long-wall method. Without going into all the details of mine layout, I can describe the method as one in which, after a face 100 meters or so in length is prepared, the coal removal operation moves back and forth across that face, totally extracting the panel as the operations proceed. The roof is supported only in the immediate vicinity of the face, and as the face recedes, the roof collapses behind the working. The support system might use wooden props or, as a more modern method, metal supports. Metal supports can be hydraulically powered and can also be designed to move themselves forward by a hydraulic mechanism. When a whole set of such mechanized supports is arranged in a line along the face, they constitute a kind of movable shield. The central elements in long-wall technology are removing the coal from the face, getting it to the haulway, and controlling the roof. The progression of technologies in this kind of coal mining has been from hand operations with picks, to mechanical undercutting and blasting the coal off the face, to the use of continuous mining machines (*combines* is the Soviet term) that move across the face, tearing off the coal and loading it onto a face conveyor. Progress in roof control has moved from wooden pit props to metal supports, and ultimately to the creation of hydraulically powered, self-advancing mechanical supports.

The final step in the evolution of this technology is the introduction of complexes of equipment, in which transport out of the working area is by a conveyor running parallel to the face, on the frame of which moves a coal cutting device. Behind the conveyor frame is a line of hydraulically operated roof supports, attached to the frame of the conveyor. As each pass is made across the face, the whole complex (including a flexible conveyor) is moved forward into position for the next pass.

The first generation of combines introduced in the USSR after World War II were "wide-web," which, as they passed across the face, removed a section a meter or more thick. These have now largely been supplanted by "narrow-web" combines that make a much shallower cut (0.6–1 meters into the face). An alternative kind of equipment used for long-wall mining is the scraper or plow, but scrapers have been used only slightly in the USSR—in 1973 only 3 percent of underground coal was so produced (*Ugol'*, 1974:2, p. 24).

The technological challenge is to design equipment that will work, to differentiate it for variations in thickness and pitch of seam, strength of coal, character of the roof, and so on. The major preoccupation of Soviet coal mining R and D has been with this kind of equipment, and the following section reviews the Soviet history of development with an eye to understanding and evaluating the innovation process.

Combines

The development of continuous mining machines began in the USSR in the early postwar period with the creation of the Donbass-1 wideweb combine. The experimental model was created in 1948 by Dongiprouglemash. Series production began in January 1949, with the first lot of 50 being prepared in the first two months of the year. Total output in 1949 was 255 machines, and another 400 were produced in 1950. Thus, there was rather rapid mastery of this machine; it is said to have been based on an earlier cutting machine, the MV-60, and there was a high degree of parts commonality. The Donbass-1 was described in 1957 by an official of the Joy Manufacturing Corporation as the "best of the Soviet continuous mining machines, a very good machine, and a machine of their own design and invention" (U.S. Congress: House, 1957). Criticisms of it by its Soviet users are somewhat less flattering— they say that it was not reliable, that it wore out rapidly and was difficult to repair. It was also suited to only a fairly narrow range of the conditions found in Soviet mines.

The Donbass-1 became the major wide-web combine used in Soviet mines—by January 1959, there were 1300 of them at work (Bratchenko, 1960, p. 83) —this was more than a third of all combines on hand. A considerable effort went into improving it over the period when it was being produced, mainly in the direction of increasing its power and in adapting it to differences in seam thickness and coal strength. Some five modified models were developed to handle thicker seams, to deal with viscous coal, and to work from the frame of a conveyor. In 1959, the original Donbass-1 model was taken out of production and in the early sixties there was fairly rapid replacement with various modifications (especially the LGD-2), all of the wide-web type.

The potential advantages of narrow-web combines were already well understood in the fifties, and in the USSR, as in Western Europe, they were being experimented with (*Ugol'*, 1964:11, p. 11). The short advance required in each cycle with narrow-web combines makes it possible for the mechanical roof supports to bridge the whole distance between the face and the line of roof fall, and in particular to eliminate props in the space where the combine and the conveyor work. The short advance also facilitates employment of flexible conveyors, so that it is not necessary to disassemble the conveyor for each move. The narrow-web miner moves faster across the face, so that more cycles are accomplished in a given period of time. Also, it seems that it is easier for narrow-web combines to work by the shuttle method (i.e., back and forth), so that they do not have to be moved and set up again at the opposite end of the face after each cycle, an operation that took a lot of time with the older combines. In short, narrow-web combines have been the main technical direction for increasing productivity in long-wall mining. Even more important, they were the precondition for the development of self-advancing complexes.

The transition to narrow-web machines took place rather rapidly in Western Europe. A Soviet writer reports that already by 1957 over half of all underground coal output was mined by narrow-web technology in West Germany, France, Belgium, and Holland (*Ugol'*, 1960:10, p. 63). The Soviet coal industry, however, was rather slow in making this transition to narrow-web machines, and as a corollary, to the use of complexes. The USSR began development work on narrow-web combines at more or less the same time as in Western Europe and apparently also at the time envisaged moving from wide-web to narrow-web technology. An experimental prototype of such a combine was developed at the Malakovskii experimental plant and was tested in a mine in 1958. After some modifications it began to be produced under the designation K-52M and began to replace the Donbass-1 combine in several mines beginning in 1959 (*Ugol'*, 1960:7, pp. 18–20; and 1962:8, pp. 2–4). This miner was fairly quickly and successfully integrated into a complex—the OMKT—in the Moscow basin. That complex is described as "the pride of Soviet coal miners" and its developers were awarded a Lenin Prize in 1961 (Mel'nikov, 1968b, p. 137). A second narrow-web model—the DU-1—was produced and intended for rapid diffusion in the Donbass mines. It was planned to have 500 of them at work by 1958 (*Ugol'*, 1960:2, p. 13). But the rate of progress in moving to narrow-web technology in the industry as a whole, and in the Donbass especially, was slow. In 1964, 616 wide-web combines were produced compared to only 355 narrow-web ones (*Ugol'*, 1965:3, p. 1), and it was only at the end of the sixties that narrow-web combines came to account for half the stock (*Ugol'*, 1969:4, p. 76).

One of the first of the new models (the DU-1) was apparently poorly conceived to take advantage of narrow-web potential, and this led to a controversy over the advantage of the narrow-web principle in general. An engineer whose views (I would judge) reflected those of the miners set off a debate with an article in the February 1960 issue of *Ugol'*, arguing that in practice the narrow-web approach was less advantageous than the wide-web and suggesting that the latter be retained as the main direction of technical advance. To judge from the responses in the journal (in issues 7 and 10 in 1960 and issues 2 and 3 of 1961), one would gather that his was an eccentric view—most of the respondents repudiated his argument and held that the advantages of narrow-web technology were clear. But there must have been a stronger support for his view than appeared on the surface. Several years later it was said that the slow progress toward the creation of complexes was explained by the obstinate but futile efforts of the design institutes in the Ukraine to develop new wide-web models and the appropriate kind of mechanized support systems to work with them in complexes (*Ugol'*, 1964:4, p. 12). Another comment suggests that the NII of the branch had been very dilatory in creating specifications for the design of new narrow-web combines (*Ugol'*, 1965:3, p. 6), which may indicate that there were still some serious doubts on the part of the technological decision makers about narrow-web machines.

One of the interesting features of that debate was that the proponents of narrow-web technology generally referred to the experience of Western countries as demonstrating its high productivity and technical advantages. As with most controversies in the Soviet system, we can find in the literature only limited information about the nature of the dispute, but it may be that this is another example of the USSR outgrowing the effects of earlier technical isolation, and that it took a while to reorient the technical lines that had grown up in this isolation. Whatever conservatism there may have been on the part of the users and the R and D organs, it could only have been reinforced by the inertia of the machinery producers. For well-known reasons they much preferred to keep on producing familar models rather than introducing new ones.

By the mid-sixties the competitive advantage of narrow-web combines was no longer an issue (an all-Union conference in 1966 on generalizing experience with the new equipment underlined that this was the most important technical direction for developing the coal industry in the Seven Year Plan—*Ugol'*, 1966:10, p. 70), but the development of the complex in which to employ them, including proper conveyors and self-advancing supports, still took quite a bit of time. Apart from the OMKT complex already mentioned (which was limited in its application primarily to the Moscow basin), the most important of the narrow-web complexes was the KM-87. It is considered one of the successes of Soviet coal mining R and D, and its developers, too, were awarded a state prize in 1968 (Smekhov, 1970, p. 3). It is worth recounting the development history of this complex in some detail because it gives us some insight into what kind of systemic difficulties have slowed the creation and diffusion of new equipment in Soviet underground mining.

The KM-87 Complex

Giprouglemash first started work on this complex in 1957. The first 12 experimental sections of the M-87 hydraulic support, intended for use in a complex, had been produced and tested by 1960. The combine for it was the K-52M narrow-web combine described earlier. The KM-87 complex was first put together in the form of a single prototype produced at the Malakovskii factory in 1960 and sent for testing in the Stalin mine of the Lugansk regional economic council. After a couple of months of tests (24 November 1960 to 29 January 1961) it was concluded that an experimental lot (*opytnaia partiia*) should be produced (Shcherban', 1969, vol. 2, pp. 236–238). This task was assigned to the Toretskii plant, which was supposed to produce seven units of this complex in 1960–1961, but produced none. Of the eight units specified in the plan for 1962, only four were produced (*Ugol'*, 1963:6, p. 5). Apparently the experimental lot was finally completed by the end of 1963, and finally, by the first quarter of 1964, five complexes were put into mines for testing (*Ugol'*, 1964:8, pp. 54–55). The tests revealed many defects in the design, such as leakages in the hydraulic system, unreliable linkage to the conveyor, and inadequate strength in the roof bars. But the experimental use was considered sufficiently encouraging that it was decided to produce a larger batch of 20 in 1964, with some improvements (*Ugol'*, 1964:8, p. 55), though apparently only four of that intended batch of 20 were produced (*Ugol'*, 1975:5, p. 26).

Concurrent with these delays, Giprouglemash was experimenting with adaptations for thicker seams and sloping seams, and some of these modifications may have been incorporated in the models produced for testing. In any case, by the end of 1964 the test program was essentially complete, and a decision was made to begin series production of the KM-87 complex.

The Toretskii plant was supposed to focus all its attention in 1965 on series productions of M-100 and M-87 hydraulic supports (*Ugol'*, 1965:3, p. 4), but 1965 was mostly a year of slippage; only in 1966 did the KM-87 begin to be produced in significant numbers. Even then plans were badly underfulfilled—although output was planned at 140 units in 1965 and 200 in 1966 (*Ugol'*, July, 1965, p. 38), by 1 January 1968, there were only 82 complexes at work, and by 1 January 1969, 147. In 1968 the total amount of coal produced with the help of this equipment was 20.1 MT, or only 4.5 percent of all underground output (Smekhov, 1970, p. 219). In short, it took about 11 years to develop this complex and get it working to a significant degree.

For widespread adoption, the original complex had also to be prepared in many modifications to deal with variations in conditions of slope, kinds of coal, and different seam thicknesses. Without going into the details, each of these modifications involved a similar history of slippages and delays.

Despite the delays in development, the KM-87 is a success in the sense that it works, and has been effective in raising productivity. I would conclude that the designers seem to have handled their part of the task well. Also, once the machine was actually available, mine management was eager to adopt it. The villains of the piece, the ones responsible for the long lag between design and commercial diffusion, seem to be the machinery plants.

The variation in conditions in Soviet mines means that, even with modifications, the KM-87 complex can be used at only a small fraction of all the active faces, and to do for the other cases what the KM-87 did for some, it is necessary to develop many additional types of combines, supports, conveyors, and other items. As an indication of how many cases are involved, one source says that 13 types of mechanized supports and complexes, eight types of narrow-web combines, three scrapers, and nine types of movable conveyors are being produced in series. In addition, industrial prototypes exist for 15 more types of mechanized support, 14 more narrow-web combines, and 12 conveyors (*Ugol'*, 1974:2, p. 24). In this broad range of equipment types the successes have been fairly rare in relation to the number of failures. When we look at this broader range of cases, the design organizations do not always come off so well, and it is less plausible to assert that responsibility for failures rests primarily on the machinery plants.

One author describes the experience with the MK-1 combine: Giprouglemash designed the MK-1 combine in 1958, prototypes were produced and testing showed the machine to be unsuitable. In 1964, a new model was designed and a lot of 22 was produced, but this model was again found ineffective. In 1967, a third model, put out in 70 units, suffered the same fate. The author concluded, "Over this ten year period about 5 million rubles were spent on the creation of these three machines, and the coal industry still has not gotten the machine it needs" (Koshkarev, 1972, p. 153). He then goes on to say that this same institute (Giprouglemash in Donetsk) has worked on developing equipment for thin seams for many years but has still not created the needed equipment. The Minister of the Ukrainian Coal Ministry says that DonUgi also has worked on this problem for four years at a cost of 1.2 million rubles and has not yet produced the machine (*Izvestiia*, 2 April 1975).

On the whole, however, I believe that the greatest weakness in the system is the problem of *producing* the new models that the design organizations develop. It seems often to be accepted that the basic designs are suitable, the prototypes have been tested and accepted, but the production plants have failed to execute them in reliable versions or have neglected to produce them in the quantities needed (see *Ugol'*, 1967:2, p. 33). As one source puts it, the technology for dealing with flat and gently sloping seams of thin to average thickness is perfected; all that is holding them back is "the small output and unsatisfactory quality of the corresponding equipment" (*Ugol'*, 1967:1, p. 17). Another complains that "domestic mechanized supports, combines, and other equipment sometimes lag behind foreign models in the quality of manufacture of particular assemblies and parts, which significantly affect their reliability and durability . . . . Domestic combines mine 130–150 thousand tons before capital repair, foreign ones 400–500 thousand tons (Iatskov, 1976, p. 18). The difficulty is often the failure of other branches and plants to supply adequate components and materials. One of the biggest problems with the narrow-web machines and plows is in the quality of the chain. These machines work by gripping, and moving along, a chain stretched across the face, and Soviet industry seems to have been simply unable to provide chains of the requisite quality. Chains for the scraper conveyors are also a problem. Two solutions (both of which the Soviet Union is now using) are to either import that one crucial item from abroad or drop the chain method in favor of cables that pull the machine across the face.

The slow advance of new underground mining technology seems to be a classic illustration of the unresponsiveness of the suppliers to the needs of the client due to "departmental barriers" characteristic of the Soviet economy. Coal-mining machinery has been produced in the Ministry of Heavy, Energy, and Transport Machine Building (Mintiazhmash), for which coal-mining machinery is a minor sideline. Mintiazhmash habitually underfulfills the production targets for coal-mining machinery and fails to add the capacity the investment plan calls for. There is poor communication between its production plants and the R and D design institutes in the coal industry responsible for developing new models of machines. For a while it was thought that the problem could be solved by having the Ministry of the Coal Industry control the R and D funds for coal-mining machinery development. The funds were allocated to Minugol', which then contracted for development work by design bureaus in Mintiazhmash (*Resheniia Partii i pravitel'stva po khoziaistvennym voprosam*, vol. 8, M, 1972, p. 531). This seems not to have worked either, and in 1973, jurisdiction over the coal-mining machine plants themselves were transferred to Minugol'. We may properly be skeptical that this shift will do much to improve performance. Two years after this shift had taken place, B. F. Bratchenko, the Minister of the Coal Industry stated:

> Much remains to be straightened out in machine building for the coal industry. . . . Granted, production here is already being converted to the output of equipment needed by the miners, and combines for operation on thin and steep seams are being put into production. Nonetheless, the work of the All-Union Coal Machinery Association cannot be termed satisfactory. Managerial mistakes have led to interruptions in the production process and a slowdown in the rate of output. [*Izvestiia*, 13 December 1975]

Obviously many of the incentive problems that inhibit innovation in Soviet industry generally can be expected to continue even when the coal mines and the producers of coal-mining machinery have a common ministerial boss.

EQUIPMENT FOR STRIP MINING

One of the principal means for raising productivity in the coal industry is to change the proportions in favor of strip mining, which has a strong advantage in capital cost per unit capacity and in labor productivity (cf. Tables 4-1 and 4-2). As for other advantages, safety is increased, and the time required to put a facility into production is shorter.

Given the existence of appropriate reserves, in locations where the coal can be shipped to market economically, the advantage of strip over underground mining is largely a function of the progress of stripping technology, which primarily has taken the form of larger equipment. Increases in the unit size of equipment have reduced the cost of removing overburden, thus moving the frontier of competition progressively against underground mining. Unfortunately, the need to expand coking coal output has limited the shift from underground mining in the USSR, since Soviet reserves of coking coal that are strippable are relatively small. Furthermore, most coal suitable for strip mining is in the East, though in the European USSR the Dnepr and Moscow basins, both producing lignite, have some strippable resources. In both the latter cases the beds lie relatively horizontally, and the layer of overburden is not terribly thick. Two large mines in the Moscow basin use large draglines to shift the overburden to the worked out area. In the Dnepr fields the overburden is soft enough to strip with bucket-wheel excavators, with transport by bridge conveyors. In the East, the situation is geologically more complicated; faulted and sloping seams are common. In such cases the mines must be excavated progressively deeper and wider, with the spoil moved outside the mine. A notable exception in the East is the Kansk-Achinsk basin, which has a thin overburden and a flat, thick coal seam.

Since the mid-fifties, the Russians have increased underground output mostly in order to meet the growing demand for coking coal and in a few cases have increased underground output of steam coal on the basis of the locational advantage of particular mines. Otherwise, the policy has been to meet incremental needs for coal by strip mining. During the 8th Five Year Plan, 65–67 percent of the output increment was to come from strip mines, though in actual fulfillment this figure turned out to be only about 56 percent. During the 9th Five Year Plan, the share of strip-mined coal in both the planned and realized increment was about 68 percent, and during the 10th Five Year Plan the share will be somewhat lower—52–53 percent according to the plan targets. The problem is the difficulty in finding ways of using the huge strip mine potential of the eastern regions. Several of the associated R and D problems involved are discussed in Chapter 7, and the discussion here will be limited to the technology of strip mining itself.

Productivity in strip mining, and hence its competitiveness, can be increased by improving equipment. Among the choices to be made in designing the production scheme for a strip mine are the "system" to be used and the size of the mine. The major systems that the Russians distinguish are: (1) "the transportless system," in which overburden is removed to the spoil area by dragline excavators; (2) the transport system, using either trucks or rail transport for moving overburden and coal; and (3) removal of overburden to the spoil bank by conveyors. (Hydraulic mining is another possibility but is insignificant in the total.) Whatever system is chosen, there must then be decisions about the kinds of equipment to be used—i.e., between mechanical shovels, draglines, or bucket-wheel excavators for excavating overburden and coal; and between conveyor, truck, or railroad haulage. For each kind of equipment, decisions must be made about size and such technical parameters as the power characteristics of transport equipment, given the average and ruling gradients, distances, and so on. Finally, all these different elements must be coordinated in a system-optimizing manner. Every mine is more or less a special case, and a detailed critique of technological decisions would have to analyze each mine or type of mine separately.* Lacking the expertise to go into the subject at that level, I shall try instead to develop what I believe are some general conclusions about the decisions that have been made in designing the projects and equipment for strip mining.

First, it seems fairly well agreed that too little emphasis has been given to the transportless method, the share of which in stripping work has remained more or less steady at around a third. A very thorough survey done at the beginning of the Seven Year Plan acknowledges the transportless method as the cheapest and says that the industry has tried to expand its use as "domestic plants have mastered production of the basic equipment" (Zasiadko, 1959, pp. 264–265). Another author specifically states that, had bigger draglines been available, economic advantage would have called for more use of the transportless system (Loginov, 1971).

Second, Soviet mine designers have apparently underrated the potential of trucks as an alternative to railroad haulage of overburden and coal. The share of trucks in hauling overburden was long almost imperceptible, though it is now beginning to creep up to nearly a fifth. Even for coal, trucks account for only about a third of the total moved. Bigger trucks and more robust design would have enhanced their competitiveness compared to rail transport. As one author says, "the currently held notion as to the limited applicability of trucks in open-pit mining of coal is explained primarily by the lack of large capacity haulage equipment which would correspond to a rational system of coal mining" (*Ugol'*, 1966:5, p. 24). Another explains further that,

---

*Another distinctive feature of the technological decision-making process is that the equipment used tends to be produced in relatively small numbers. Differential design to meet the needs of a particular situation and the fact that each of these units of equipment is an item of very high productivity mean that the numbers of units of any particular model that is produced will be relatively small. We may be talking in terms of a few hundred trucks, with excavators and draglines in the tens. In the case of bucket-wheel excavators, each machine may be essentially unique. More will be said later about Soviet efforts to achieve economies and to meet performance specifications by setting up a *tipazh* that envisages both differentiation to cover all situations and design commonality that simplifies design and production efforts.

"the high cost of truck transport in existing mines is explained by the relatively small haulage capacities of the trucks used, and by the rapid wearing out of engines and tires, which necessitates very large outlays for repair" (ANSSSR, 1968, p. 140).

In both cases there was a failure to take advantage of the savings inherent in large unit sizes for equipment. Soviet strip mines for coal tend to be very large, so that they could economically use very high capacity equipment. In the United States, the nature of the market, the size of the deposits, and transport considerations keep numerous relatively small producers competitive (Christenson, 1962). The average output per strip mine in the USSR is over 3 million tons per year, whereas the average in the United States is only a little over 100 thousand tons per year. For a benchmark to suggest what size equipment would be economically suitable for the Soviet industry or some notion of the relevant current technical level in the United States, we should thus take as the U.S. analogue the relatively small fraction of U.S. mines that are in the same size class as the Soviet ones.

It is not completely clear whether the failure to take advantage of larger equipment represents a mistake on the part of those who design the open-pit operations or a constraint imposed by the inability of the supplying industries to create the equipment. My interpretation is that, although there may be some biases on the customer side, the problem is largely on the supply side. Soviet mining experts seem to be well aware of the advantages of larger equipment but have not been able to get it as soon as they wanted it or in the quantities they want. The best way to clarify this question is to look at the development history of particular kinds of equipment.

Excavators

The history of excavators shows a struggle to get larger models produced and a persistent lag behind what the Soviet coal industry wants, what would seem to be technically and economically feasible, and what is being supplied to mines in other countries. A brief history of the development of various excavator models will be helpful to show the sequence in which various sizes were produced and the delays in getting any given size produced.

Consider first mechanical shovels. Before the Second World War, the Russians had no large excavating equipment, and this is one of the main reasons they did almost no strip mining (in 1940, less than 4 percent of output was from strip mines). During the Second World War the USSR acquired a number of foreign shovels (presumably under Lend-Lease) larger than any they had themselves produced, which became the basis for considerable expansion of strip mining in the Ural region (Mel'nikov, 1957, p. 20) *. The first domestic model, developed after the Second World War, was the SE-3, a shovel with a 3-meter bucket, produced in the Ural Heavy Machine Building Plant (*Uralmashzavod*) during 1947–1956. A large number of these were produced —the thousandth one had been produced by the end of 1954 (Rozenfel'd, 1961, p. 446). One author, discussing strategy for new models at a later date, attributes the success of the SE-3 to the fact that, by specifying what it most needed in the form of one model, the coal industry was successful in focussing the attention of the producing plants on producing that model in large numbers (*Ugol'*, 1970:8, pp. 47–49). It was given a 4-meter bucket in some applications, and after 1956 essentially the same shovel, but with more improvements, was produced under the designation EKG-4† (Mel'nikov, 1958, p. 21). At some later date further modifications were made to produce a model with a somewhat larger bucket—the EKG-4.6. Soviet writers seem to agree that this was a very successful model. Mel'nikov says it was the best in the world in its time (Mel'nikov, 1966, p. 104).

The next step up was the EKG-8 fitted with either a 6- or an 8-meter bucket in different versions. The design work on this shovel was done by the Izhorsk plant, beginning in 1955 (Mel'nikov, 1968, p. 55), and it was produced from 1957 by Uralmashzavod. By the end of 1964, 30 of these machines were in operation, and by the end of 1966, 76 (*Ugol'*, 1967:11, pp. 20–21). Another model, the EVG-6, was adapted for stripping work from the basic EKG-8 design.

As of 1970, the three shovels so far described (SE-3, EKG-4-4.6, EKG (EVG)-8) accounted for almost two-thirds of all mechanical shovels used in strip mining coal. Soviet coal industry commentators were aware that this put them at a big disadvantage compared to other countries where a much greater range of shovels were available (e.g., see Kuznetsov, 1971, p. 279).

Experience with producing larger shovels has been much less im-

*Mel'nikov, incidentally, is probably the most noted Soviet authority on strip mining. He is an Academician but has also had executive responsibility in the coal-mining industry.

†In these designations, E stands for excavator, K indicates shovels intended for removing the mineral, V for removing overburden. G indicates mounting on caterpillar tracks. The number indicates bucket size in cubic meters, and when a second number follows a hyphen, it indicates boom length.

pressive, either in terms of getting models produced or ensuring that they justify themselves in operation. The next increase in size involved one model with a 12.5-meter bucket and another with a 15-meter bucket. In 1967 the Council of Ministers passed a special decree to encourage development of the Kuzbass coal fields, for which it was decided larger shovels were needed ("O merakh neotlozhnoi pomoshchi po dal'nei-shemu razvitii ugol'noi promyshlennosti Kuznetskogo bassenia" in *Resheniia Partii i pravitel'stva po khoziaistvennym voprosam*, vol. 6, pp. 393–401). The decree called for production of three prototypes of the EKG-12.5 by the Izhorsk plant, though apparently in the end only one prototype was produced. The decree envisaged series production of this model by 1969, but series production did not actually begin until about 1975. For this model, therefore, there was a five-year slippage in the production time table. Design work on a 15-meter shovel, the EGL-15, began as early as 1948, the decision to produce a prototype was taken in 1949, and it is claimed that a prototype was produced in 1950 (ANSSSR, 1968, p. 105). But another source says the first of these shovels began to work only in 1958 (*Ugol'*, 1960:3, p. 8). There were only three in operation at the end of 1970, which suggests that nothing more than prototypes were ever produced (Zhuravlev, 1971, p. 7). The design of this shovel must have been unsatisfactory in some way. Minugol' decided to stop using the one they had at Cheremkhovo, though it is not explained what was wrong with it (Ministerstvo ugol'noi promyshlennosti, 1969, p. 40). Mel'nikov, in describing the *tipazh* for shovels as envisaged for the foreseeable future did not include a 15-meter shovel (*Planovoe khoziaistvo*, 1974:8, p. 80).

The largest mechanical shovel attempted so far is a 35-cubic meter shovel, originally intended to be produced in 1959 (Zasiadko, 1959, pp. 267–268). A prototype was developed in 1960, which was put to work in 1965 (ANSSSR, 1968, p. 108) at Cheremkhovo in the Irkutsk basin. It must not have worked very well, and apparently underwent a long series of trials and modifications. A statement in 1964 indicated that series production would begin in 1966 (*Ugol'*, 1964:11, p. 74). This goal was not met, and it was finally retargeted for series production sometime in 1971–1975. So in this case the slippage was something like 10 years.

Excavating shovels can be made very large. There is a 138-cubic meter machine in use in the United States (*Mining Engineering*, October, 1974, p. 4), and Soviet planners have long talked about much larger shovels. For example, it is thought that a 100-cubic meter shovel would be appropriate for overburden removal at some of the Kansk-Achinsk mines (*Ugol'*, 1967:4, p. 43). The Novo-Kramatorsk plant was said in 1967 to have started design of an EVG-100/70 (*Ugol'*, 1967:11, p. 21) and Mel'nikov includes it in the *tipazh* of overburden strippers still intended for development (*Planovoe khoziaistvo*, 1974: 10, p. 80). There is a controversy as to the desirability of so large a shovel. One Soviet engineer surveying the history of excavators says that experience has shown that for stripping work the advantage is clearly with draglines and that it would be a mistake to spend resources on developing a new series of large shovels (*Ugol'*, 1970:8, pp. 47–49).

The story of dragline excavators of the walking type is also one of slow upward movement to larger models, with many delays and setbacks along the way. Soviet production started with the ESh-1 model, which has a 3.4-meter bucket and a 38-meter boom. It is interesting that the initiative for the development of this excavator was taken by the coal industry—the prototypes were developed in repair plants under the control of the coal industry rather than in heavy-machinery plants. Production of the industrial version, however, was handled by the Novo-Kramatorsk plant in 1949–1952 (Dombrovskii, 1969, p. 59). During 1952–1957 it was replaced with the ESh-4/40 produced at the Novo-Kramatorsk plant, one of the two plants that have produced almost all excavators.* This excavator underwent various modifications to 5/40 and 5/45 versions, which utilized more fully the capacity of the basic design.

The same plant also produced another family of models designated as the 6/60, 6/80, and 8/60 (*Ugol'*, 1967:11, pp. 20–21).† The first of

---

*In these model designations, ESh stands for walking excavator (*ekskavator shagaiushchii*), and the two numbers show the size of the bucket in cubic meters (before the slash) and the length of the boom in meters.

†A basic feature of dragline design is a trade-off between bucket size and boom length. The longer the boom, the smaller the weight that can be worked at the end of it. Shortening the boom permits the exacavator to be fitted with a larger bucket. The principle of producing several modifications of a given machine on this basis has been followed in Soviet excavator production. According to N. N. Mel'nikov, the *tipazh* of excavators worked out by Tsentrogiproshakht envisaged two such modifications for each basic model (*Ugol'*, 1967:4, p. 43). Mel'nikov also gives a somewhat different version of the optimal *tipazh*, in which only some of the draglines have more than one alternative version (*Planovoe khoziaistvo*, 1974:8, p. 80). There seems to be a dispute on the optimal combination of boom length and bucket size. Apparently Soviet design has preferred long booms. The biggest excavator has an 80-meter bucket on a 150-meter boom, whereas the biggest U.S. excavator has a 168-meter bucket on a 94-meter boom (*Bituminous Coal Facts*,

these machines began work in 1959, and series production began in 1961 (ANSSSR, 1968, p. 106). A still later generation, consisting of the 10/60 and 10/70, was reported as undergoing industrial testing in 1964 (*Ugol'*, 1964:11, p. 74), though other sources suggest testing did not take place until 1965 or 1966 (*Ugol'*, 1967:4, pp. 43–44). Actual production of the 10/70, however, was assigned to Uralmashzavod (the other plant capable of producing excavators) along with a couple of modifications—10/75A and 13/50. The lot of the 13/50 for industrial testing (*opytno-promyshlennaia partiia*) was to be produced in the first half of 1967, but so far as I can tell it was never actually produced (see Zhuravlev, 1971, p. 7). This is in spite of the fact that it was thought about 35 or 40 would be needed, and that it would be put in series production in 1969–1970 (*Ugol'*, 1967:4, pp. 44, 48). I have been unable to find an explanation of what went wrong on that model.

Uralmashzavod also developed a succession of models in its own R and D institutes. The first of these was the 14/65 family, on which design work started in 1948 and which was produced in 20 units during the years 1949–1957 (Rosenfel'd, 1961, p. 43). The initial model was produced in record time, probably because of the high priority of the Volga-Don Canal for which it was originally intended. Subcontracting was an important element in the achievement, and as an interesting parallel to what we have described as happening in the electric power equipment case, it is said that many of the elements were produced and delivered before the final design was settled (ANSSSR, 1968, p. 100). The prototype that went to work on the Volga-Don Canal in 1950 was a 14/65 version; the series version produced from 1952 on was a 14/75 model (*Bol'shaia Sovetskaia Entsiklopediia*, 2nd ed., vol. 48, p. 405; and Dombrovskii, 1969, pp. 60–61). This excavator was redesigned and modernized in 1957 to become essentially a new model, the 15/90 excavator. The first of these was tested in 1959 (ANSSSR, 1968, p. 108). A 20/65 version, only one of which was ever produced (*Ugol'*, 1967:4, p. 43), was a modification that sacrificed boom length to allow a bigger bucket. All of these larger excavators turned out by Uralmashzavod were produced in rather small numbers;

1972). Some Soviet authors characterize the U.S. trend as "progressive" and one that the USSR should follow. One possible interpretation is that the length is set by the minimum needed to cut a wide swath and that the designers have then had to be content with a small bucket because relatively low-strength steel for the boom limits the weight that can be handled (see Mel'nikov in *Ugol'*, 1976:2, p. 42 and Loginov, 1971, pp. 190–191; for more on steel, see below).

there were only 45 of them in operation as of the end of 1970 (Zhurav-lev, 1971, p. 7).

A significant step upward at Uralmashzavod was to a 25/100 model. It is claimed that a 25-meter model was "prepared" in 1957 (Rosen-fel'd, 1961, p. 443), and a 1964 source says that an experimental model "has been accepted" by a State Commission (Ugol', 1964:11, p. 74). Since the 1967 decree for the development of the Kuzbass mentioned above stipulated that two 25/100 "prototypes" would be produced for testing, one in 1969, and one in 1970, the earlier effort must have been very tentative. Apparently one prototype was actually produced in response to the 1967 decree and was in operation by 1970 (Zhuravlev, 1971, p. 7). I have seen no figures as to how many of these machines were then produced during the 9th Five Year Plan, but apparently not many. Mention is made of the project at Uralmashzavod to build a 50-cubic meter walking excavator (Rosenfel'd, 1961, p. 447), but noth-ing ever came of this.

Finally, the biggest dragline the Russians have so far attempted is the ESh-100/100. These very large excavators are intended for work in the new mines to be constructed in the Kansk-Achinsk basin. This excavator was originally intended to be a 80/100 model, and the *tekhnicheskii proekt** for this excavator was said in one source to have been completed in 1967 (Ugol', 1967:1, p. 3) and in another source, in 1964 (ANSSSR, 1968, p. 60). According to the 9th Five Year Plan, a prototype was to be in operation and series production to be organ-ized by the end of the period—i.e., by 1975. For some reason this de-sign was modified to carry a 100-cubic meter bucket instead of the original 80-cubic meter bucket, and a prototype had been assembled and moved to the work area in the Nazarovo mine in the Kansk-Achinsk basin early in 1977 (Ugol', 1978:1, p. 27).

I am in no position to evaluate the technical qualities of these ex-cavators nor have I been able to find an evaluation by any Western expert. N. G. Dombrovskii, the principal Soviet authority on excava-tors, says that the Soviet models compare very favorably with foreign products, though in the early years they were excessively heavy. He also claims that some of their design features are original and advan-tageous, mentioning especially the trussed-mast boom, and the hydrau-lic walking mechanism (Dombrovskii, 1969, pp. 60–61). Not all experts

---

*Soviet design procedure usually goes through three stages of increasing detail and seriousness. The first is the *avanproekt*, followed by an *eskiznyi proekt*, and then *tekhnicheskii proekt*, which is the actual engineering design.

agree, however, and one spokesman for the Soviet coal industry criticizes both these features specifically (*Ugol'*, 1964:10). But the most important conclusion to be drawn from this development history is that, although in the early postwar years Soviet excavators seem to have matched U.S. machines in size, the industry subsequently has lagged seriously in developing satisfactory models of larger sizes.

Finally, the Russians have long been interested in large capacity bucket-wheel excavators. As of 1975 they had 30 or so at work in various mines. The longest experience seems to be in the Dnepr mines, where soft overburden, lignite, and relatively mild weather create good conditions for use of such excavators. But the excavators so far developed have inadequate force on the cutting edge of the bucket to permit their use either for overburden or coal in most mines without blasting. Also, they do not work properly in winter because the frozen ground is too hard to cut and because the conveyors that are part of the system break down. The planners, however, would like to employ bucket-wheel excavators in those other situations and have been experimenting for a long time.

In the development of wheel excavators, again the initiative came from the coal industry to the extent that the first one was an essentially homemade model produced at Ekibastuz in the mid-sixties, by modifying an EKG-4.6 shovel. In 1963 the Donetsk plant *imeni* LSKM created a relatively small prototype, the ERG-350, which was put into experimental operation at the Irsha-Borodinsk mine in East Siberia. On the basis of the results of testing this prototype, a modification, the ERG-400-D, with a capacity of 1,000 tons per hour was prepared in 1966 to be used in the Ekibastuz mine, and it was in experimental operation by September 1966. The tests revealed that it lacked sufficient biting force to break up the coal without blasting, it did not work well in cold weather, and its motors lacked adequate capacity. Many modifications were made in the testing process, and after a year's experimental operation it was accepted in September 1967 by a state commission and recommended for series production. Out of these experiments there also arose plans for a new model, ERP-1250, with a capacity of 1250 cubic meters per hour, and this is the kind of machine most widely used.

At some point the Russians became interested in the bucket-wheel excavators used in the East German open-pit lignite mines. One of these with a capacity of 1,000 tons per hour (the Sr_s K-470 model) was installed at Ekibastuz in 1969 and was in operation in 1970–1971. At some time also the Germans supplied an SR_s K-2000 machine with a capacity of 3,000 tons per hour. This is one of the few cases of technological transfer in the coal industry and one of the few cases from Eastern Europe in any energy sector.

But the Russians apparently intended to use these German machines only as models to be copied and have gone on to produce domestic models with capacities of 3,000 and 5,000 tons per hour. The Novo-Kramatorsk plant was supposed to produce a larger model, the ERG-1600, with an hourly capacity of 3,000 cubic meters per hour (ANSSSR, 1968, p. 62), and according to Zhuravlev (1971, p. 8), there was one such machine working in 1970. I believe an accurate statement of the situation is that bucket-wheel excavators have become fairly important in coal extraction but still have not been effectively mastered for the removal of overburden.

Trucks

Soviet decision makers in open-pit mining have generally underrated the potential of trucks, and there is a kind of tacit assumption that, when the transport method of removing overburden is used, the work will be done by railroad. It is instructive, for instance, that the 1967 decree mentioned earlier on strengthening the equipment base for strip mining concerned itself with every other kind of equipment including locomotives and dumpcars, but did not mention trucks. Given rail transport for overburden, it often follows naturally that coal will also be moved by rail, especially since much steam coal is not treated and is shipped directly to users in cars loaded directly in the mine. This method is used at Ekibastuz, for example. The emphasis on rail transport grew in part out of the fact that for a long time there were no high-capacity trucks that the mine designers could build into a system. A need for such trucks was recognized early, but the R and D effort to develop them has been unsuccessful. The commitment to rail seems all the more dubious when it is realized that for a long time the technological level of mine railroad transport was relatively low. Until the sixties, coal mine haulage used steam engines almost exclusively, and only in the 9th Five Year Plan did a fairly decisive shift to electric and diesel traction take place. Most Soviet dumpcars until the 9th Five Year Plan period had capacities smaller than the trucks used in big U.S. mines.

Nor is this interpretation merely the projection of an outsider. More recently a much greater appreciation of the potential of trucks has arisen in the USSR, especially for the flexibility they permit in mine operation. As one Soviet writer says, "without setting ourselves the goal of critiquing the decisions of projectmakers [to use rail transport] . . . it can be affirmed with certainty that in many cases these decisions do not meet the sharply rising economic demands" (*Ugol'*, 1971:1, p. 40).

For a long time only relatively small trucks were available to the industry. The model most widely used was the 25-ton MAZ-525, produced in the Minsk automobile factory beginning in the fifties. The MAZ-525 was supplemented by the smaller IaKZ-210-E dump trucks produced by the Iaroslav plant with a 10-ton capacity and the KrAz-256 with a capacity of 10–12 tons, which was apparently produced from the early sixties (*Ugol'*, 1967:11, p. 24).

These models were modified somewhat over the years, sometimes by fitting a larger bed to fully utilize the truck's capacity when material of lower density was hauled. For example, the MAZ-525 was modified into a new model called the MAZ-530, which had a 30-ton capacity. Later, apparently this same truck was fitted with a larger engine and was produced in a version with tandem rear axles that gave it a capacity of 40 tons. An important limitation on all these trucks was that they were basically general-purpose dump trucks intended for use in heavy construction; their structural strength and power were not optimized for the condition of strip mining. Coal is less dense than many other cargoes so that truck capacity is underutilized with coal, and too much of the work goes into hauling the weight of the truck itself (*Ugol'*, 1972:3, p. 30).

Primary responsibility for designing and producing heavy trucks for open-pit mining has rested with the BelAZ plant in Zhodino, Belorussia. The BelAZ designers, together with the coal industry institutes, worked out a *tipazh* of trucks that was supposed to provide something suitable for every situation. It envisaged a series of models up to 180 tons in capacity, with various kinds of transmissions, dumping systems, and wheel arrangements (*Ugol'*, 1972:3, pp. 30–34). Other sources speak of plans for 300 and 500 ton vehicles as well. (The notion of a *tipazh* is fundamental in Soviet R and D practice and more will be said about it below.) The smallest of the basic models in the BelAZ *tipazh*, and still the most widely used, is the BelAZ-540, a 27-ton dump truck produced also in a large-body version to haul less dense material. The BelAZ-540 was first produced in 1961 (*Ugol'*, 1963:2, p. 21) and was put into series production in 1967 (Kuznetsov, 1971, p. 294).

Progress in producing models larger than 27 tons has been extremely slow. A 40-ton truck had been a high priority of the mines for a long time. A review of technical needs made at the beginning of the Seven Year Plan (i.e., in 1959) said such a truck was an urgent necessity (Zasiadko, 1959, p. 276). A prototype of the 40-ton version of the BelAZ-548 was built in 1962 (*Ugol'*, 1963:2, p. 21). This model seems to have been produced in small numbers by 1970, and began to be used in coal mines during the 9th Five Year Plan. Thus the delay between the articulation of the need and actual availability in this case was 10 to 15 years.

The mining industry has been eagerly awaiting long-promised models with capacities of 65 and 120 tons. The timing is not very clear, but already in 1960 the coal industry was counting on using the 65-ton truck. A prototype was produced quite early—there is a picture of a prototype in a 1968 book (Mel'nikov, 1968, p. 180). Presumably this truck was tested and accepted, since, according to an article in *Ugol'* (1976:2, p. 7), preparations were made during the 9th Five Year Plan (1971–1975) to begin series production of it. But progress was obviously not very rapid, since a deputy minister of the coal industry says in 1976 only that the process of mastering series production (*osvoenie seriinogo proizvodstva*) of the 65-ton model has been started (*Ugol'*, 1976:2, p. 7). Soviet statements compound several terms for beginning in a way that makes it uncertain whether they have actually begun any activity, and through 1978 I have not seen any statement that this model is actually being produced in series.

The guidelines for the 10th Five Year Plan specifically mention starting production of 75-ton and 120-ton haulage vehicles. A model described as the BelAZ-549, with a capacity of 75 tons, is pictured in a 1976 newspaper article and is described as "the firstborn of an industrial lot" (*promyshlennaia partiia*). This formulation suggests that this model has already been tested and is in production, and a coal industry source corroborates that the industry began to receive these vehicles in 1977 (*Ugol'*, 1978:2, p. 37).

A prototype of the 120-ton model is supposed to have undergone industrial testing as early as 1972 (Mel'nikov, 1972, p. 85). The 9th Five Year Plan specified that it was to be "mastered" during the plan period—i.e., "industrial production" was supposed to have been achieved by 1975 (Baibakov, 1972, p. 130). Obviously it was not, since that goal has been restated for the 10th Five Year Plan in the *Guidelines* for that plan. As late as 1978 it is still described as undergoing testing (*Ugol'*, 1978:2, p. 37). This is the first model to be based on electric motors on individual wheels; all the earlier models have used hydraulic transmissions.

Another idea which was highly touted at one point but which has not yet reached successful development is the *trolleivoz*, a concept for a vehicle with a diesel engine generating electricity to be used by motors on the wheels but also equipped with apparatus for drawing power from a network. For movement on its regular route over heavy grades out of deep pits, the vehicle would draw power from a contact net, but its independent power source would permit it to maneuver on its own in the mining areas. This approach would also economize on energy input through use of regenerative braking as the empty vehicle returned to the lower levels. The BelAZ plant was supposed to produce this 65-ton capacity vehicle in 1967, and an experimental batch was supposed to be produced in 1968, to be tested at mines. Not much progress has been made on this idea; a 1971 source said that there had still not been industrial testing of a prototype (Loginov, 1971, pp. 198–199), but it may not yet have been abandoned—Loginov says that it holds great promise for the future.

In any case, as of the mid-seventies the largest vehicle the mining industry had been able to obtain had a capacity of only 40 tons. This seems to be an example of the situation common in Soviet industry, in which the designers and producers charged with the production of new equipment simply can't deliver it, even though it is not really radical or especially demanding in its technology. Moreover, foreign models have demonstrated the feasibility of all the basic technical features. According to *Bituminous Coal Facts* (1972), the U.S. industry uses vehicles with capacities of 150 tons, and apparently some 220 ton vehicles were also in use by the mid-seventies. The time it takes from the first design effort to a commercially producible and usable product is extremely long. The general interpretation for this kind of innovational failure in the Soviet economy (and it occurs frequently) emphasizes such problems as inability to get crucial inputs or failure of the responsible organization to support the project in a crunch when it competes with other objectives. Another common element in many of the explanations is that the Soviet system has all the pieces but just can't fit them all together. Most commentators have been inclined to think that Soviet design organizations are competent and employ well-trained and talented personnel. I am beginning to think, however, that there may be something wrong in that area as well—i.e., the designers too often produce a design that they ought to know can't be produced or won't work under the conditions to which it will be subjected.

The delay in increasing vehicle size and the fact that there are no really large vehicles means a serious disproportion in size between excavators and trucks. The 27-ton truck is not big enough for use with the EKG-8 excavator, which as explained earlier is the main excavator currently in use for removing coal. There is apparently a rule of thumb that a truck should be loaded with three or four excavator bucketfuls. When trucks are too small, the capacity of the excavators is not fully realized, and when the excavators are too small, the trucks spend too much time waiting instead of hauling. The existence of this disproportion is a commonplace in the Soviet literature, and the coal industry experts are unhappy about it (see especially *Ugol'*, 1965:11, p. 28; and *Ugol'*, 1972:3, p. 30).

Rail Transport Equipment

Rail transport remains the most important method of hauling overburden to the spoil dump—in 1970, trucks hauled only a sixth of the overburden moved by the transport method (Zhuravlev, 1971, p. 10), and the share could not have risen much since. Technological progress in this area is again mostly a function of increases in size of equipment and of a change from steam traction to diesel and electric traction. The main features of Soviet technological history here are the relatively late shift from steam locomotives and a fairly steady upward creep in the size and productivity of the train units. There were some electric locomotives already at the beginning of the fifties, but as late as the end of the decade, steam traction still accounted for over 60 percent of the rail transport work (Zasiadko, 1959, p. 276). Proportions began to shift rapidly in the sixties, and by 1966 the share of *electric* traction was 60 percent (*Ugol'*, 1967:8, p. 55); it must be much higher now. The power of locomotives has gone from a tractive effort of about 80 tons in the fifties to 240 tons in the new diesel units that began to be introduced in the seventies. Dumpcars were mostly of 40–60 tons capacity in the fifties, but by the mid-seventies about half of them were in the class of 100–105 tons (*Ugol'*, 1976:2, p. 37). With this larger capacity equipment, train productivities doubled between the fifties and the seventies (Mel'nikov, 1957, p. 27; and N. P. Zhuravlev, 1971, p. 9). The contrast between this picture of success in getting larger rail equipment and the delays in raising capacities of other types of equipment may be explained partly in terms of help from the Eastern Europeans. The larger dumpcars are imported from Poland and Czechoslovakia, as well as being domestically produced, and one of the larger locomotive units used in open-pit mining (which combines an electric control locomotive with a diesel unit permitting operation away from a power supply) was developed in East Germany (ANSSSR, 1968, pp. 42–43).

Auxiliary Equipment

The technological level of strip mining is decisively affected by the excavating and haulage equipment discussed so far, but a variety of auxiliary equipment is also important. Several Soviet commentators note that Soviet strip mining uses relatively few bulldozers, scrapers, or wheeled loaders, though these are common in the U.S. industry (cf. Table 4-2). The problem again is that the equipment-producing industries will not supply the kind of equipment suited to strip mining. Major complaints are that the available bulldozers are insufficiently powerful to meet the demands of strip mining and that engines and tracks break down. The most powerful dozer is that based on the DET-250 tractor, with a 300 HP engine, while elsewhere in the world capacities up to 500 HP are available (Loginov, 1971, p. 149—actually, Caterpillar advertises a 700-HP dozer in Soviet journals).

Another auxiliary process that has been neglected is auger extraction. The usual role for augers in connection with strip mining is to clean out exposed seams on the fringes of an excavation where the overburden has become too thick to remove economically (Carroll Christenson, *op. cit.*). The 8th Five Year Plan (1966–1970) envisaged series production of augurs (ANSSSR, 1968, p. 40), but apparently nothing beyond very preliminary work was attempted. According to Mel'nikov, "thus far we have not given the requisite attention to auger mining of coal, and have so far done only some experimental work in the Kuzbass" (*Ugol'*, 1971:8, p. 43). Little progress was made in the 9th Five Year Plan (1971–1975), since five years later the only novel thing he has to report is that a prototype auger produced by the Donetsk plant was given industrial trials in 1975 (*Ugol'*, 1976:2, p. 38).

About 85 percent of all overburden removed in Soviet strip mines must be blasted before excavation (Mel'nikov, 1972); drilling equipment therefore has an important effect on productivity. The drilling of the holes for these charges is a labor-intensive job that takes about 12–13 percent of all Soviet labor in strip mining (Dobva, 1973, p. 143). Technical progress in this area has involved going from percussion drilling to rotary drilling, and, within the latter, from scraper bits to roller bits and combined scraper and roller bits. Accompanying this change is a shift to larger diameter holes and greater flexibility in the angle at which the holes are drilled. These changes in turn make for more effective blasting, in some cases moving a good part of the blasted rock all the way to the dumping area. In general outline, the Soviet adoption of these innovations has been as follows:

Improvements in drilling blast holes for strip mining came rather late—the Russians were slow to move away from cable-tool percussion drilling. They even did a lot of hand drilling until the fifties, and the share of cable-tool percussion drilling stayed at a fifth to a fourth up to the middle of the sixties. A commentator at the beginning of the seventies pinpointed drilling as a weak element because of dependence on obsolete cable-tool drills. This required a lot of heavy physical labor, the quality of the driving steel was bad, and reliability was poor so that the equipment was idle a great deal of the time (V. P. Loginov, 1971, pp. 138–139). Things began to change in the second half of the sixties, when decent models of rotary drills began to be produced. The shift to these more productive models came in the 9th Five Year Plan period (*Ugol'*, 1974:11), though by 1975 these machines still constituted only about a third of the stock (*Ugol'*, 1976:2, p. 35).

There is quite a bit of information on the development histories of the various models of drills, and one of the interesting aspects of the development process is that there seems to have been some elements of design competition between different design institutions, each working with its own experimental plant. Rather than going through these details, however, I simply offer my general conclusion that the situation with these drills must have been much like that with the other equipment. The industry knew what it needed just by following U.S. experience, and the research institutes designed the corresponding models. The new designs did not get turned into prototypes very fast and, when they did, contained defects. When these were ironed out and the machines put into production, they were still not produced quickly enough, and so at any given time the industry has been burdened with a great deal of obsolete equipment.

CONCLUSIONS

One clear and consistent theme has run through this chapter—the difficulty experienced by the coal industry in getting the equipment that it needs to meet its goals of modernization and productivity growth. The kind of explanation suggested has in general been consistent with what the general literature on innovation in the Soviet economy has to say. The Soviet economy is a seller's market; lateral communications for making the consumer's wishes known and giving them leverage are weak; the incentive system that guides the producers of equipment inhibits their interest in producing new models. The explanation usually runs in terms of the behavior of the production enterprise, but the coal industry example adds the useful reminder that the lack of responsiveness may involve higher levels as well. Important in the charge against the sector producing coal mining machinery is that it has failed to add the capacity to produce the new designs. This is a decision made at the ministerial level, where the pressure of other demands has led the decision makers to use the capital investment resources intended for this purpose in other directions. On each of the occasions when the Council of Ministers has taken up the issue of how to get the coal industry reequipped, its decrees have made a special point of directing Mintiazhmash to increase the capacities of the coal machinery plants or to "realize" fully the investments earmarked for that purpose. (See, for instance, the 1968 decree on the technical reequipping of the coal industry in *Resheniia partii i pravitel'stva po khoziaistvennym voprosam*, vol. 7, pp. 64–73; and the similar decree of 1973, in volume 9 of the same source, pp. 485–490). Similarly, in excavator production, the ministry seems to feel stronger pressures to fulfill other parts of its production program than to produce the needed excavators. The plants producing excavators are primarily concerned with producing steel mill equipment—it is the ability of these plants to produce very large metal parts that fits them for the production of excavators. But excavators are only a minor sideline in their total product mix. Moreover, parts of the excavator projects are farmed out to a lot of other plants, for which, again, this activity is only a sideline (Loginov, 1971). There is a difference here from the situation in electric power—the plants serving coal mining typically produce for a variety of customers and so are in a better position to behave as if they were in a seller's market.

Clearly evident in the coal-mining machinery industry are the familiar environmental conditions that make enterprises reluctant to innovate, especially the difficulty of getting components and materials. The problems with chains for conveyors was mentioned earlier, and there was a similar problem with material for conveyor belts. It seems that steel quality has been one of the major problems constraining design and production of better excavators. One source says that dragline excavators cannot be improved until steel with higher yield strength is available to the producers. Whereas the steel most commonly used in American excavators is said to have a yield strength of 70 kg/mm^2 down to temperatures of $-50°$C, Soviet excavator producers must use steel with a yield strength of 40 kg/mm^2 down to $-40°$C (*Ugol'*, 1974: 3, p. 30). Another author says that if better steel were available it would be possible to reduce the weight of the boom and make it possible to carry a larger dipper (*Ugol'*, 1976:2, p. 42).

Beyond this confirmation of generally accepted interpretations, however, the coal industry also offers some insights to this producer-customer interaction at the R and D stages that precede actual production, and I would like to conclude with some discussion of this question. This interaction can be broken down into two stages: the specification of a range of different models in some area (i.e., the *tipazh* mentioned earlier) and the process of design within the specifications of this *tipazh*.

The *tipazh* approach seems to be a distinctive feature of the Soviet system. The objective is to develop a catalog of all possible applications of some kind of equipment and then establish specifications for a set of models that is to cover these needs. An analogous phenomenon is very common in the market economy, though it is more likely to be a creation of either the buyer (as in military procurement) or the seller, rather than something agreed on jointly by all buyers and sellers. This stage would seem to be a quintessentially cooperative job that must absorb the perspectives both of the producers and the users of the equipment. One might think that the buyer is in the best position to specify what is needed, but it is probably more realistic to assume that the user will often be somewhat conservative and may need some stimulation from the production side. It is sometimes claimed for example, with respect to a U.S. analogue, that the military has a dangerous tendency to overspecify what it wants and is better off when considerable latitude is left to the production side to use its R and D capability to generate better possibilities. The production side should in any case be allowed a considerable input so that the specifications reflect its intimate awareness of what is possible in terms of producibility, though, of course, it too can be technically conservative and may need to have its imagination stretched.

I believe that for many kinds of coal machinery the initiative in setting specifications is to a large extent in the hands of the coal industry. The sources seem generally to indicate that the initiative in setting up the *tipazh* for excavators has been the responsibility of Giprotsentroshakht, though this job also involves the cooperation of customers using open-pit methods in other kinds of mining as well (ANSSSR, 1968, p. 57). I believe, however, that the producers have had some influence on this *tipazh*. I suspect the producers have a lot to say regarding what is feasible, and they always have a strong interest in ensuring a high degree of commonality in parts and subassemblies. The excavator producers have big KBs that do technical economic studies that we can find in the literature, and we have a considerable volume of writings from chief designers and other people on the industry side about their input into this process. Moreover, certain characteristic basic design traditions exhibited in the excavator models all along seem to reflect considerations imposed by the production side—e.g., the relatively small bucket size for any given boom length mentioned earlier.

There is also a *tipazh* for trucks, and there are more indications that in this case the coal industry has had inadequate input. Most discussions speak of the *tipazh* as the creation of the BelAZ designers, but more telling are some complaints on the part of coal industry spokesmen that it does not reflect coal industry needs. One author says that the BelAZ models are deliberately patterned after American equipment and that this is inappropriate since Soviet open-pit mines tend to work deeper horizons than American mines. The greater depth of working in Soviet mines means a need for more climbing and hence a higher power/weight ratio than is characteristic of American equipment (Vilenskii, 1962).

For underground equipment, the literature does not speak of an explicit *tipazh*, and the definition of needs and setting of specifications is more likely to be done in terms of individual machines. For underground mining, the comprehensive view would in any case seem to need to be oriented more to the problem of compatibility among different kinds of machines, rather than compatibility among different models of a given kind of machine. In the development process for coal-mining machinery there is excessive concentration on central items of equipment, with inadequate attention given to balanced and integrated R and D work on all the elements (similar to the problem noted in Chapter 3 for the electric power industry). One critic says that in underground mining the development of equipment is carried out piecemeal; not enough consideration is directed to where labor can be saved in the whole cycle (*Ugol'*, 1976:11, p. 60). We have already noted the dominant focus on excavators in open-pit mining, with too little attention given to equipment for such auxiliary processes as road building, cleaning up, and leveling.

Specifications for equipment, whether individually or in the form of some sort of overall *tipazh*, are still very general, and what kind of equipment is actually developed will be determined at the actual design stage. The same basic interactive, cooperative desideratum is important at this level. Here we are getting closer to themes that are well developed in the general literature on Soviet R and D, but the coal industry suggests a couple of new twists.

First, much depends on who controls the R and D institutes, and the conditions of production may not leave a great deal of room for choice in this respect. In the case of excavators and trucks, the KBs are closely identified with the production side. As mentioned earlier, the first Soviet walking dragline was produced by a plant in the coal industry itself, as was the first bucket-wheel excavator. But, for such complex and heavy machinery, serious design and prototype production has to be done by the producers. The production of excavators has been concentrated in four large heavy-machine-building plants, and each of these has a large KB for excavator design. In the 1962 decree on improving open-pit mining, one of the measures taken was to greatly enlarge and strengthen these KBs ("O merakh po dal'neishomu razvitiiu i sovershenstvovaniiu dobychi poleznykh iskopaemykh otkrytym sposobom," *Resheniia Partii i pravitel'stva po khoziaistvennym voprosam*, vol. 5, Moscow, 1968). Under these conditions, no matter what the specifications say, the designs produced are likely to be dominated by producer considerations rather than by user needs. The Soviet economy seems to inspire two kinds of biases here. First, the producers seem to have a great interest in commonality of parts between successive generations, different sizes, and model variations. Adherence to this principle is very explicitly laid out both for excavators (as in books like Dombrovskii, 1969) and for trucks (Dronov and Shatokhinaia, 1970, pp. 91–98). This principle seems almost certain to inhibit adequate adaptation to different conditions and to account for things like the small intermodel steps in expanding excavator capacity.*

*A detailed study of Soviet locomotive design strongly suggested that this design principle was a considerable inhibition to progress (William Boncher, "Innovation

Furthermore, there is a great tendency toward concentration on the production side, limiting the possibility that different producers will come up with different design ideas. There are only four plants that have played any significant role in producing large excavators, and in the Soviet system even these four do not really have design independence. As mentioned earlier, one of the families of excavators developed at the Kramatorsk plant was assigned to Uralmashzavod for production. For underground mining machinery, there seems to be a little more diversity based on the distinction between regional basins. It should be remembered that, in the shift to narrow-web combines and to complexes, the Moscow basin group of institutes and producers were the first to take this direction, and in doing so, provided an alternative to contrary views in the Donbass group.

---

and Technical Adaptation in the Russian Economy: The Growth in Unit Power of the Russian Mainline Freight Locomotive," Ph.D. Dissertation, Indiana University, 1976.

5

Nuclear Power

Thus far we have looked at the performance on technologies in a historical way. In all the cases we have studied, the USSR was usually catching up with technologies developed elsewhere, so that the path to be followed was clear. It was a matter of organizing to get something made and mastered in use, usually following the experience and even copying the specific equipment already developed by someone else. To the extent that there was technology transfer, it was through technology embodied in equipment or in the form of learning to do something that others were already doing. But the USSR is also engaged in R and D in some frontier areas, where the path to follow is not so clear, and Soviet R and D decision makers are more on their own, as in nuclear power and several other novel energy sources—solar power, geothermal, and others. Because in these areas the USSR is working at somewhere near the same stage of technological development as Western countries, examination of their approaches and experience may hold lessons for Western countries. These examples may be more informative in illuminating how the USSR handles such problems as moving from speculative basic research to more applied development, how R and D planners decide when a development is ripe for undertaking large investment commitments, and other aspects of the commercialization process. This chapter deals with the nuclear program, and the following chapter with several more exotic cases.

CURRENT STATUS
OF NUCLEAR POWER

As indicated in an earlier chapter, the Soviet Union began its effort to develop nuclear power quite early. It claims to have created the first nuclear power plant in the world (the Obninsk station), though some Western sources dispute this on the grounds that it was essentially only a test reactor. They have also had a fairly large R and D program—lots of resources, high priority, a variety of programs and approaches. The growth of nuclear power capacity, however, has been relatively slow in the USSR, as shown in Table 5-1. Table 5-2 shows the major stations in operation and under construction or planned.

TABLE 5-1. Growth of Nuclear Power Capacity and Output

End of Year	Capacity (*Million KW*)	Output (*BKWH*)
1970	0.9	3.5
1971	1.37	3.8
1972	2.77	7.1
1973	3.20	11.7
1974	3.70	18.0
1975	4.70	20.2
1976	6.00	25.0
1977	7.30	34.0
1978	8.30	45.7
1980	(18.50)	(82.8)

SOURCE: Campbell, 1979, pp. 16, 18; *Atomnaia energia*, 1979:4 p. 279; Nekrasov and Pervukhin, 1977, pp. 6, 110, 114.

Several different kinds of reactors have been used. The original Obninsk station used a water-cooled, graphite-moderated reactor, and two others among the early installations—the Troitsk station and the Beloiarsk station—were also of this type. The Beloiarsk station was distinguished by having a superheat cycle. The main commercialization effort, however, was made with a pressurized water reactor (described by the Russian initials VVER), which was first installed in the Novo-Voronezh station in the Ukraine. The first unit at that station was a prototype reactor of 210 MW capacity, commissioned in 1964. As the station was expanded, successively larger versions were employed, and the version that became the standard and that was produced in series was the VVER-440, with 440 MW electrical capacity. The first unit of this model was commissioned in 1971. This reactor was also used for the Kola station in the USSR (completed in 1973) and was the model exported to the Eastern European countries and to Finland. Stations using the VVER-440 are in operation in East Germany, Czechoslovakia, and Bulgaria as of 1977 and are being built or have been contracted for in Hungary, Poland, and Cuba. Finland has now purchased its third VVER-440. This reactor is now being scaled up, and the next unit to be installed at the Novo-Voronezh station will have a capacity of 1,000 MW.

TABLE 5-2. Soviet Nuclear Power Reactors in Operation and Under Construction

Station Name or Location	Reactor Name or Number	Type of Reactor	Capacity (MW)	Operation Begun or Intended
Siberian (Troitsk)	—	LWGR	6 × 100	1958–1962
Beloiarsk (near Sverdlovsk)	No. 1	LWGR	100	1964
	No. 2	LWGR	200	1968
	No. 3	FBR	600	1980
Novo-Voronezh	No. 1	VVER	265	1964
	No. 2	VVER	365	1969
	No. 3	VVER	440	1971
	No. 4	VVER	440	1972
	No. 5	VVER	1,000	1979
Dmitrovgrad	VK-50	BWR	50	1965
	BOR-60	FBR	60	1969
Shevchenko	BN-350	FBR	150	1973
Bilibino	ATETs	LWGR	4 × 12	1973–1976
Kola (near Murmansk)	No. 1	VVER	440	1973
	No. 2	VVER	440	1974
	No. 3b	VVER	440	1980
Leningrad (at Sosnovyi Bor)	No. 1	RBMK	1,000	1975
	No. 2	RBMK	1,000	1976
	No. 3	RBMK	1,000	1980
	No. 4	RBMK	1,000	1980

Kursk (at Kurchatov)	No. 1	RBMK	1,000	1976 (Dec.)
	No. 2	RBMK	1,000	1979
	No. 3[b]	RBMK	1,000	1980
Armenian (at Metsamor near Erevan)	No. 1	VVER	440	1976 (Dec.)
	No. 2	VVER	440	1979
Chernobyl (near Kiev)	No. 1	RBMK	1,000	1977
	No. 2	RBMK	1,000	1978
	No. 3[b]	RBMK	1,000	1980
Smolensk	No. 1	RBMK	1,000	1980
	No. 2	RBMK	1,000	(?)
South Ukrainian (at Nikolaev)	No. 1[b]	VVER	1,000	1980
West Ukrainian (at Rovno)	No. 1[a]	VVER	440	1979
	No. 2[a]	VVER	440	(?)
	No. 3	VVER	1,000	(?)
Kalinin	No. 1	VVER	1,000	(?)
Ignalina (Lithuanian SSR)	No. 1	RBMK	1,500	1981–1982

BWR = Boiling-water Reactor; FBR = Fast Breeder Reactor; LWGR = Light-water Graphite Reactor; RBMK = Channel-type Light Water Graphite Reactor; VVER = Pressurized-light-water Reactor. See text for fuller descriptions.
[a] *Atomnaia energiia*, 1976:2.
[b] *Atomnaia energiia*, 1979:4, p. 279.

But at some point in this expansion, a decision was made to commercialize a different kind of reactor as well—the graphite-moderated-type, called in its current version the RBMK-1000, the first of which was installed in a station near Leningrad. In the plan for the future, such reactors will play an increasingly important role (More will be said about the competition between the two types below). The Leningrad reactor had a capacity of 1,000 MW, and most of the stations now being built use this same size reactor. Those farther into the future will use larger units. A 1,500 MW reactor of the RBMK type is under development for the Ignalina station to be built in Lithuania, and this is expected to be an intermediate step on the way to reactors with 2–2.4 GW capacity.

The Soviet Union is also one of the world leaders in the creation of breeder reactors; as of the mid-seventies one experimental reactor was in operation, and a larger, second-generation version (600 MW capacity) was under construction.

The explanation for the relatively late start and slow growth of nuclear power in the USSR is the competition from cheaper alternative energy sources. Given the easy availability of fuel and its relatively low cost, fossil-fired power stations appeared more economical. The situation began to change at the end of the sixties, when the successful operation of several commercial nuclear plants provided encouraging evidence on capital costs per unit of capacity, and when the difficulty of expanding fuel output and the high cost of fuel (at first on a regional basis) made nuclear plants look comparatively more attractive. So during the 9th Five Year Plan period (1971–1975) a shift to nuclear power was begun, especially for the European region of the country—some new stations were brought into operation, and construction was started on several more. In the 10th Five Year Plan (1976–1980), 13 to 15 Megawatts, or about 20 percent of all new generating capacity, is to be accounted for by nuclear plants. A considerable number of new plants will also be started to ensure continued growth in subsequent periods.

The Soviet nuclear power program certainly seems to be a success—it has commercialized two types of reactor and is on the way to making nuclear power provide a significant fraction of total electric power. The Soviet position today is that the cost of power from nuclear plants is competitive with that from conventional plants and that operating experience shows adequate reliability and safety and demonstrates the viability of the basic technological decisions taken. A few questions about this picture will be introduced later, but it seems to be essentially correct. In short, the USSR has a working, domestically developed technology for fission and is among the world leaders in the two major frontier nuclear technologies of breeder and fusion reactors. The rest of this chapter will explore how the Soviet system has approached this area of R and D, explain what is distinctive about the decisions that have been made on such questions as reactor types and safety, and examine what position the USSR is taking on some major controversial issues, such as the role of the breeder reactor, the reprocessing of waste fuel, and the international diffusion of nuclear technology.

STRATEGIC CONSIDERATIONS IN NUCLEAR POLICY

There is a standard genre of article in the Soviet energy literature dealing with "possible paths of development of nuclear power" or "the future of nuclear power," which is helpful in understanding the strategic considerations that have guided nuclear power decision making and influence current decisions. And it may be useful to emphasize here that virtually all decision making in nuclear power contains a large element of R and D decision making. The technology is changing just enough that every decision must include a big R and D effort; each choice involves questions to which the answers are uncertain. These discussions usually turn around several main themes.

One is the issue of fuel supply for nuclear power—the potential reserves of nuclear fuel and the cost of producing it—which has important implications for many issues in nuclear R and D. The long-range supply picture for fissionable material is an area of controversy in the USSR, though it is usually discussed in guarded terms. A. M. Petrosiants, Chief of the State Committee for the Peaceful Uses of Atomic Energy, has introduced a new chapter in the 3rd edition of his book on Soviet nuclear policy entitled "Are the Supplies of Uranium Large?" (Petrosiants, 1976). My impression is that the USSR does not have very large domestic reserves. The USSR has depended on Eastern Europe in part, and the directives for the 10th Five Year Plan list exploration to discover more nuclear fuel as a high priority task. Though the USSR has the surplus enrichment capacity that induces it to produce nuclear fuel for export to numerous countries, such export is carried out only in the form of toll enrichment of uranium supplied

by the customer (except to Eastern Europe and possibly to Finland).
It also seems likely that Soviet leaders would want, for strategic reasons, to be independent of outside supplies, so in thinking about future supplies they are less influenced by the reserves situation in the
world as a whole and possible prices in international trade than by
domestic production potential. Petrosiants makes a special point of
this: "The development of this branch of technology in the socialist
countries does not depend in any way on supplies of uranium from the
capitalist world" (A. M. Petrosiants, 1976, p. 260). The question of
uranium supplies has important implications for nuclear R and D,
especially for the importance attached to the breeder reactor and the
timetable for its development as the transition technology bridging the
period between the era of uranium-fueled thermal reactors and the
advent of fusion sources.

A second major theme has to do with the adaptation of nuclear
reactors to supply energy for additional uses beyond electric power
generation. Under Soviet conditions the competitive status of nuclear
power is basically determined by the regional problem, and there is
thus a strong motivation to diversify the applications of nuclear power
in the regions where it is competitive, to serve more energy uses than
power generation alone. The R and D planners accept as a high priority task the development of nuclear reactors to supply high-temperature heat (either direct or by-product) for chemical and metallurgical
processing, as well as lower-temperature heat for space heating. This
latter goal continues a tradition well established in fossil-fuel power
generation.

A third factor governing R and D decisions is the relatively short
history and small scale of nuclear power production in the USSR to
date, which leads to special considerations in Soviet thinking about the
fuel cycle. Specifically, there are today relatively smaller stocks of plutonium available to the Russians in the form of spent fuel than in the
United States. It is from the reprocessing of such fuel that the first
generation of breeder reactors will be supplied with fuel. Because Soviet nuclear planning envisages fairly early introduction of the breeder
and rapid expansion once it is commercialized, the nuclear R and D
decision makers seek high breeding rates and modifications of the current thermal-neutron reactors to produce more than the usual amounts
of plutonium. The interest in using by-product heat for high-temperature processing interacts with the concern for short doubling times to

generate an interest in new coolants, such as dissociating oxides of nitrogen for a gas-cooled fast-neutron reactor.

Another issue with which nuclear policy makers must deal is size of reactors and nuclear power stations. As explained in Chapter 4, the Soviet power industry has a long-held faith in economies of scale, and this has a strong influence on nuclear power R and D decisions. Conventional stations are now projected to be 4.8–6 GW in capacity, using 800, 1,000, and 1,500 MW generating units. The largest nuclear station in 1977 is the 2 GW Leningrad station, using two RBMK–1000 reactors, but several stations using this reactor will be expanded to 4 GW, using four such units (N. A. Dollezhal' and I. Ia. Emelianov, in *Atomnaia energiia*, 1976:2, p. 122—these are the Leningrad, Chernobyl, and Smolensk stations). As already explained, the VVER reactor is being scaled up to a 1,000 MW version, and a 1,500 MW RBMK is being created for the Lithuanian station. This size goal has an important impact on R and D and nuclear policy in general. The biggest obstacle to increasing the size of VVER-type reactors is the pressure vessel used; this is one of the factors that has led to an increased priority for the modular-type RBMK. Also, the combination of large size, desire to use by-product heat, and commitment to large-scale use of the breeder combine to produce, what would seem to an outsider, an extraordinary safety hazard in the form of extremely large plants using novel technology, located in close proximity to large population concentrations. The factors that inhibit this line of development in the United States do not seem to operate in the Soviet case.

FORECASTING AND MODELING IN NUCLEAR R AND D

There is a considerable Soviet literature on policymaking approaches for the nuclear power sector, focussed on how to forecast and optimize. The approach is usually described by Soviet authors as iterative hierarchical modeling which seeks to optimize the internal composition of the nuclear power generating sector, while fitting it in an optimal way into the overall primary energy supply on the one hand, and into a concrete electrical (and ultimately heat) supply system on the other. A description of this approach may be found in a pair of articles by Academicians N. A. Dollezhal' and L. A. Melent'ev in *Vestnik ANSSSR* (1976:11, pp. 51–61; and 1977:1, pp. 87–99). Another typical example

is a book by Batov and Koriakin (1969) that analyzes economic choices
for nuclear power in a systems approach based on a rather complicated
model of the fuel cycle. The authors are clearly conversant with the
Western literature on the same subject and one of them has even published an article in the IAEA journal acknowledging the commonality
of the Soviet approach with that used in the West (*Atomic Energy
Review*, 1972, pp. 233–249).

Basic to much of the modeling is an effort to lay out the relevance
tree for electric-power-generating technologies and to forecast costs and
timetables for various new technologies. One can then use various
optimizing models that answer some questions about where the R and
D effort and emphasis ought to go. As an illustration, Figure 5-1 is
proposed as the relevance tree in Gitelman (1974, pp. 291–296). This
particular exercise has a somewhat narrow goal; the authors want to
estimate when some current nuclear plant technologies may become
obsolete, so that it will be possible to set depreciation rates to include
possible obsolescence. Depreciation is a very large item in total costs,
and cost effectiveness calculations for nuclear proposals require realistic depreciation rates. But this general approach is common to almost
all the thinking of the nuclear modelers. Another author makes forecasts of when the "commercial" form of the breeder will become available and then tries to optimize the split between conventional reactors,
converters, and breeders in new construction over the next 50 years
(Bobolovich, 1974, pp. 251–257). He makes various assumptions about
the date when the breeder will be ready, its cost relative to slow-neutron reactors, and the growth in total nuclear capacity needed. The
important consideration is that the breeder will have an economic advantage over slow-neutron reactors once it is available. At that point
it should therefore cover the whole incremental market, except that
there will not be enough plutonium to load new breeders as fast as
they could be built. Hence, there is an advantage in building converters now to stockpile plutonium to facilitate the shift in structure
of new investment when the breeder is ready. The author then inves-
tigates via his optimizing model the impact on the optimum pattern
of the date when the breeder will be ready, and whether the nuclear
power system is to work only on base load or is to have a smaller number of hours of utilization per year (which interferes with the plutonium production capacity of the breeders).

The point is that these studies seem thoroughly sensible and very
much in the spirit of analyses done in the West. They are perhaps

FIGURE 5-1. Relevance Tree for Raising the Effectiveness of Power Production

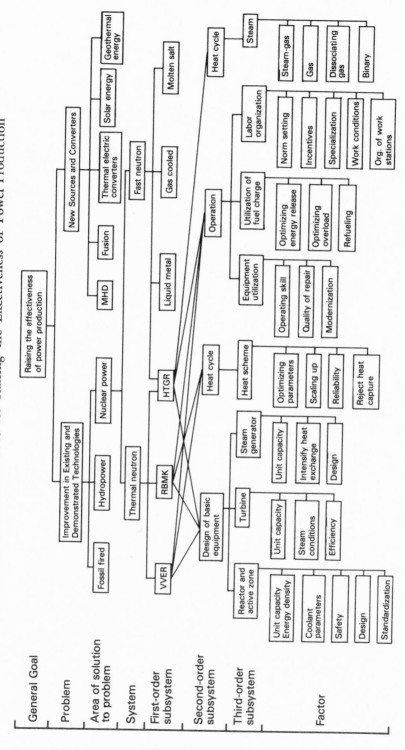

open to criticism on the same grounds. The exogenous factors (such as growth in demand or cost of construction) are unknown and are probably estimated unreliably. The number of interdependencies in a complex system of this kind is great, and the trick is to make a model simple enough to handle but complicated enough to capture the important variables and relationships. Since various models ask different questions or look at different aspects of the internal connections, sensible decision making must be eclectic, playing one model off against another. Some things that may be crucial, such as the security issues associated with diffusion of nuclear technology never get captured in the models, and decisions have to be made in the end according to less formal criteria. In short, these large models are always subject to abuse, but my impression is that the Soviet nuclear planners use them with as much understanding and economic sophistication as does anyone in the West. They probably have a distinctive set of biases, such as a desire for autonomy in fuel supply, special pessimism about the high cost and small amounts of natural uranium available, and a surprising lack of concern about safety and environmental hazards.

MAIN DIRECTIONS OF CURRENT STRATEGY

Let us consider where the Soviet nuclear R and D planners now stand on various issues in nuclear technology and their ideas about directions of R and D efforts for the future. As already explained, the Soviet nuclear power industry currently employs two main kinds of reactors—a pressurized-water reactor (the VVER) and a graphite-moderated channel-type reactor (the RBMK, which stands for *reaktor bol'shoi moshchnosti—kipiashchii*, or large capacity, boiling-water reactor). The VVER was the first to be commercialized, with the first mature version commissioned in 1971. It is said to be a reliable and economical reactor and is the one which has been offered to other countries. The RBMK was commercialized rather later (the first unit was commissioned in 1975) and represents an independent Soviet line of development. As one authority says "no other country has solved the problem" of developing a reactor of this type (N. A. Dollezhal' in *Atomnaia energiia*, 1972:2). Factors influencing the development of this reactor and its current prominent role include: long experience; an absence of novel engineering problems; modular construction, which both makes it easy to expand and susceptible to modifying and upgrading the various elements; distinct moderator and coolant, so that it is possible to choose each for its best qualities; possibility to reload without stopping; individual control over each channel; and flexibility in fuel cycle characteristics. Specifically, the last point means that it is possible to vary the amount of plutonium produced. The reactor's main deficiency is that as a boiling-water reactor it creates problems for turbine design. Relatively poor steam conditions mean a reduction in the thermal efficiency of turbines, and large volumes of steam that must be handled put a limit on turbogenerator unit size and require large cooling water supplies. The main adaptation to solve those problems is the introduction of a superheat cycle. An earlier version of this kind of reactor did employ superheating of steam, and the R and D people say that it will be relatively simple to redesign the RBMK to include superheating.

Improvements in the fuel cycle are also mentioned as important in the further evolution of this design (*Energetik*, 1974:6, p. 4). The zirconium cladding now used causes some deterioration in heat characteristics, but I think that "improvements in the fuel cycle" means basically an increase in plutonium production. The RBMK apparently also uses less thoroughly enriched uranium than does the VVER (*Atomnoi energetiki XX let*, 1974, p. 128).

The competition between these two reactor types seems to have been one of the major issues in the history of Soviet nuclear R and D decision making. It is predicted in some sources that the VVER will dominate additions to nuclear capacity over the next 30 to 40 years (*Atomnaia energiia*, 1974:3, p. 274). In contrast to that assertion, however, an official of the State Committee says in a review of a new book by Petrosiants that "it is natural that . . . among kinds of power stations, the attention is given primarily to those with channel-type reactors. The latter, as is well known [a giveaway phrase in Soviet polemics indicating what is *not* generally accepted], have made the dominant contribution to the introduction of nuclear capacity in the USSR in the 9th Five Year Plan, and this situation will be maintained in the 10th" (*Atomnaia energiia*, 1976:6, p. 503). More than 65 percent of the nuclear capacity to be commissioned in the 10th Five Year Plan will be in the form of RBMK (Nekrasov and Pervukhin, 1977, p. 47). It was originally planned to equip the Chernobyl and Smolensk stations with VVER reactors, but they were actually built with RBMK (Josef Wilczynski, "Atomic Energy for Peaceful Purposes in the Warsaw Pact Countries," and *Elektricheskie stantsii*, 1975:4, p. 14).

A recent statement by the Minister of Electric Power, P. S. Neporozhnyi suggests that there is still some argument. He says that the rationale for developing both lines is that it will be possible to compare their strengths and weaknesses and to draw more machinery plants into the task of producing the equipment (*Elektricheskie stantsii*, 1977:8, p. 3). It may be that the VVER grew out of the work on submarine propulsion, as did the basic U.S. commercial power plant designs, or was based on following the U.S. lead, so that advocates of the RBMK were able to win a role for their candidate only after a long struggle. The competitive status of the RBMK will no doubt depend in part on the requirement for the extra plutonium the RBMK can provide, which depends in part on whether the breeder is developed on schedule. (An article by Iu. I. Koriakin, et al., in *Atomnaia energiia*, 1974:4, pp. 251–256, focuses on this point.)

Breeder Reactors

The USSR may be one of the first countries to achieve successful commercialization of the breeder reactor. This will be the culmination of an R and D line followed consistently over a long period of time. The Soviet specialists have no doubts about the feasibility or desirability of the breeder. As A. M. Petrosiants says: "The basic general strategy for the development of nuclear power in the USSR is an orientation toward all-out and accelerated development and introduction of fast-neutron reactors with expanded reproduction of fuel" (Petrosiants, 1976, p. 24).

The Soviet Union is now building what is usually called a second-generation industrial plant—the BN-600 at Beloiarsk. A "first-generation industrial" plant (the BN-350) has been in operation since 1973. A British report speaks of plants in several countries, of the general size and stage of development of the BN-350, as "prototype reactors" and the next step (such as the BN-600), as "near-commercial plants" (*Annals of Nuclear Science and Engineering*, Vol. 1, 1974, p. 471).

There are numerous options in breeder reactor design and development. Such a reactor can be based on breeding fissionable U^{233} from thorium or plutonium from U^{238}. Even when this choice has been made, decisions are still open regarding such parameters as size, fuel doubling time, coolant, temperature, and pressure. Soviet efforts are oriented toward a short doubling time because the planners envisage a very large role for breeders in total nuclear capacity. The stock of fissile material invested in breeder reactors would have to grow at about the same rate as their capacity, and since it is expected that in the 1990s, the capacity of breeder reactors will double over a period of about 8 to 9 years, Soviet designers want to find a way to cut the doubling time for fissile material well below the period that is standard for Western reactor projects. The British prototype breeder at Dounreay has a doubling time of 25 years and the French Phénix, 20 years (*Science*, 30 January 1976, p. 372). The BN-600 is expected to have a doubling time of 12 years (*Elektricheskie stantsii*, 1975:10, pp. 10–11). Because it is expected that some of the plutonium produced in breeders will be employed in fission reactors adapted to special needs (such as high-temperature gas-cooled reactors for industrial processing), A. P. Aleksandrov* indicates that the goal is to achieve a doubling time in breeder of 4 to 6 years (*Atomnoi energetike XX let*, 1974, p. 211).

Reactors for Heat Production

Soviet nuclear planners have always been much interested in using the by-product heat from reactors, and in the BN-350 breeder at Shevchenko, by-product heat is used to desalinate water via a load connected to a back-pressure turbine. The R and D planners would also like to capture low-potential rejected heat (for use in urban space heating) from power plants using boiling-water and pressurized-water reactors. The Beloiarsk reactor was adapted to serve a small heat load, and the first serious experiment with such a technology is the small station at Bilibino, in the Soviet North. This station has four power blocks aggregating 48 MW and, in addition, a heat-producing capacity of 25 GKal/hr. via extraction turbines (*Atomnaia energiia*, 1973:5, pp. 299–304). But the main goal in this experiment has been to create small, simply built reactors as energy sources for remote regions rather than to add heating capacity to traditional stations, and the status of the development effort to create a prototype large-scale atomic TETs is unclear. One source says: "atomic TETsy as of the beginning of the 10th Five Year Plan are still in the beginning stage of development so that it is not possible to count on their introduction in this period" (A. M. Nekrasov and M. G. Pervukhin, *Energetika SSSR v 1976–1980 godakh*, 1977, p. 114). They add that an important task of the 10th Five Year Plan is the acceleration of the necessary research, design, and

---

*Aleksandrov has long been the Director of the Kurchatov Atomic Energy Institute and is now also the President of the USSR Academy of Sciences. His statements should be about as authoritative as it is possible to be.

project-making work . . . "in connection with which it is envisaged to begin at the end of the 10th Five Year Plan period the construction of a large-scale atomic TETs, which will lay the foundation for the further development of centralized heat supply on the basis of nuclear fuel in 1981–1990" (ibid., p. 122). A Grigoriants, head of Minenergo's Nuclear Power Administration, says such a station will be built in 1981–1985 (Pravda, 22 February 1977).

This concern flows out of the regional problem already described. The area in which nuclear power has a competitive edge is limited largely to the European part of the country, where its cost advantage flows from the high local cost of other primary energy sources. Given its appeal as a substitute for more expensive alternative primary sources, it would be useful to find ways of using nuclear power outside the base-load generating plants for which it has thus far been developed. One objective is to be able to utilize high-temperature by-product heat for chemical processing, metallurgy, and other industrial needs. The present development line in breeder reactors based on sodium as the coolant is incompatible with this goal, since the coolant leaves the reactor at a relatively low temperature (500–600°C). To meet this objective, it will be necessary to develop a high-temperature gas-cooled breeder. A great deal of experimental work has been done with various gaseous coolants, and a distinctive feature of the Soviet effort is work with dissociating oxides of nitrogen. This line of research is carried on at the Institute of Nuclear Physics of the Belorussian Academy of Sciences. An experimental reactor to try out this scheme has been proposed (Atomnaia energiia, 1974:1, pp. 11–21). According to Pavlenko and Nekrasov (1972, p. 121), enough experience has been acquired to proceed with the construction of an experimental block on this principle in the 10th Five Year Plan, though they do not say that a decision has actually been made. The creation of a gas-cooled reactor would introduce new technical difficulties, such as high pressures and risk from cooling failure in a reactor with very high energy density. The USSR has apparently never done any extensive work on high-temperature gas-cooled reactors (see below) and thus would start from a minimal base of experience.

One serious limitation on the role nuclear reactors can play, even in electric power generation, is their poor adaptation to load variations. A recent article points out that, given the load pattern and the already existing facilities, much new capacity will have to take the form of cycling and peaking units (N. A. Dollezhal' and L. A. Melent'ev, in

Vestnik ANSSSR, 1977:1, pp. 89–90). There is thus a strong desire to modify nuclear stations, both to permit them to compete with fossil fired TETs as well as to replace condensing stations, and to permit them to supply high-temperature heat.

Gas-cooled Reactors

Gas reactors apparently enjoy a high priority in Soviet nuclear R and D policy at present. Academicians Dollezhal' and Melent'ev speak of the necessity of forcing the development of high-temperature gas reactors (*Vestnik ANSSSR*, 1976:11, p. 60). There are two different reasons for taking this direction. First, given the conviction as to the desirability of fast-neutron reactors, a backup project is desirable in case the liquid metal breeder should come to grief. One source suggests that gas-cooled breeders are being thought of primarily as a backup for sodium-cooled breeders, though they might eventually have an independent appeal on grounds of cost (*Atomnaia energiia*, 1976:2, p. 135). Secondly, they want heat at high temperatures for more efficient power and for industrial processing; only a gas-cooled reactor can provide this type of heat. A gas-cooled reactor designed primarily to meet the second objective would not have to be a fast-neutron type. Thermal-neutron gas-cooled reactors, fueled with plutonium produced in the breeders in excess of what is needed to fuel new breeders, could also do the job (Meshkov, 1976, pp. 121–127).

Gas-cooled slow-neutron reactors seem to have been neglected in earlier Soviet nuclear R and D, and one wonders why they were overlooked, given a long standing interest in "atomic furnaces." I imagine that gas-cooled reactors must have simply seemed too risky—they present great problems in metallurgy, for instance. Also, the high-temperature gas reactor has been very difficult for the Western countries to perfect. In the United States, this type of reactor drove its developers into bankruptcy. More than is sometimes realized, Soviet policymakers are much impressed by the example of what Western countries are doing, and I would not be surprised that they chose for commercialization the pressurized-water reactor just because that is what was being done in the United States and avoided the HTGR because it was a poor risk in the eyes of the West. This cannot be an all-encompassing interpretation, because there is one distinctive Soviet reactor—the RBMK—but there are indications that, despite an early commitment to this type, it must have been downgraded at one point because it was different from what the Americans were doing.

Natural Uranium Reactors

The Soviet program seems never to have allocated much effort to natural uranium reactors. There is little discussion of this point in Soviet sources, but we can speculate on the reasons. The only natural uranium heavy-water reactor project has been the reactor developed by the Czechs—the A-1 reactor at Bohunice. This is a 150 MW reactor, cooled by carbon dioxide, now usually described as an experimental prototype effort (*Atomnoi energetike XX let*, 1974, p. 107). This project was undertaken by the Czechs in the first place on the strength of Soviet offers of assistance, and a Czech author even indicates that the project was proposed by the USSR (Jan Neumann, Chairman of the Czech Atomic Energy Commission, in *Ekonomicheskoe sotrudnichestvo stran-chlenov SEV*, 1972:2, p. 18). It was supposed to be a cooperative project, but in fact must have been very much under the control of the Soviet side. Soviet institutes prepared the technical design and exercised design supervision in the actual construction (*Atomnoi energetike XX let*, 1974, p. 111). It has been suggested that the Russians were not very forthcoming in supplying the promised aid and for a long time in effect blocked this development (Polach, 1968, pp. 831–851). The original hope was that the station would be finished by 1960. That was certainly unjustified optimism, but a subsequently reset deadline of 1968 was also badly missed, and the plant first operated at the end of December 1972 (*Atomnoi energetike XX let*, 1974, p. 109). It seems reasonable to suppose that the Czechs were interested in a reactor type that would make them independent of Soviet enrichment processes, but that the Russians were ambivalent about this. The Russians surely want to keep the external fuel cycle under their own control and would be reluctant to see the Czechs develop an independent enrichment capability as a part of their domestic nuclear technology. A natural uranium reactor would remove the temptation for the Czechs to develop such technology. On the other hand, Soviet opposition to a Czech-developed natural uranium technology would be understandable, both in terms of the commercial motive of wanting to eliminate competition in supplying power reactors to the other Eastern European countries, and in terms of permitting the Czechs a way to use their own uranium directly without sending it to the Russians for enrichment.

In any case, the current attitude seems to be that the A-1 experiment was not especially encouraging for further development, though one Soviet source says that it can be "described as a serious success of Soviet and Czechoslovak science and technology," apparently mostly in terms of generating experimental data (*Atomnoi energetike XX let*, 1974, p. 113). The Czechs apparently have no intention of building another one and in 1970 signed a contract for Soviet VVER-440 reactors for two stations (Neumann, 1972, p. 18).

The USSR also helped the GDR in the creation of a nuclear power plant, water-moderated and cooled, using natural and lightly enriched uranium. Subsequently, we have indications the Russians are still interested in natural uranium reactors and are doing some work along this line. The twentieth anniversary symposium at Obninsk included a paper on such reactors (*Atomnaia energiia*, 1974:3, p. 274).

FUSION POWER

One of the greatest puzzles in interpreting Soviet nuclear policy is how Soviet nuclear planners evaluate the relationship of the fission and fusion forms of nuclear power. In the United States, it is often assumed that fission should be considered only an interim technology until fusion power is perfected. Most discussions center on the question of how long the interim period will be, which in turn will influence how badly a breeder reactor is needed to stretch uranium resources. I have never seen the question of the desirability of the breeder put this way in Soviet discussions; their decisions about the breeder seem to have been taken independently of any forecast concerning the likely mastery or timetable of fusion. A. P. Aleksandrov has said that the Kurchatov Institute "has made a fundamental contribution to the ideology of developing and economically optimizing the structure of nuclear power in the coming period extending over several centuries" (*Vestnik ANSSSR*, 1975:9, p. 6). This is an obvious reference to the modeling and forecasting studies described earlier, and some additional interpretation of what he means comes from *Atomnoi energetike XX let* (1974, pp. 209–210), in which he says that, with the breeder, nuclear resources will be adequate for hundreds of years and that "such a development path was marked out many years ago in the Soviet Union." I have never seen any reference in the concrete forecasting and modeling exercises as to how fusion fits in with this concept of the breeder's role. The relevance tree study cited earlier just disposes of it by saying it is far enough in the future that it is irrelevant to the problem at issue. Aleksandrov mentions that fusion may be available by the end of the century, but mentions it only in passing and says that if it does get perfected, then *in connection with the breeder* it will be possible to "rebuild" the world's energy resources (*Atomnoi energetike XX let*, 1974, p. 213). This would seem to be a rather poorly reasoned idea.

Academicians Dollezhal' and Melent'ev, in the article cited earlier, describe the general strategy of nuclear policy, including roles and timetables for the different kinds of reactors, but never mention fusion until the last paragraph, when they say that work on such problems of the breeder as transport security, storage, and reprocessing should be combined with study of fusion at the most promising energy approach (*Vestnik ANSSSR*, 1976:11, p. 61). These two authors give a possible answer to the puzzle of why fusion is ignored as a competitor to the breeder by asserting that, although strategic decisions may have consequences several decades into the future, the uncertainties multiply so fast toward the end of such a forecast period that forecasts must be limited to about 30 years (ibid., p. 54).

At the same time, the USSR does have a large and active fusion program, and its managers seem to encourage positive thinking about the likelihood of success. A. P. Aleksandrov frequently asserts that a breakthrough is likely soon. Academician E. Velikov of the Kurchatov Institute says, in an article in *Kommunist* (1976:1), that great progress is being made and there will be a fusion power plant before the end of the century.

The current status of Soviet forecasts and priorities in fusion research is unclear but policy seems somewhat unsettled, as in the United States. The USSR has for a long time had a strong program based on the tokamak, which of course is a Soviet creation. In this line of research the last machine built was the Tokamak-10, described as the last of the purely experimental fusion installations by E. Velikov, Deputy Director of the Kurchatov Institute (*Kommunist*, 1976:1, p. 67–70). He thinks that the data accumulated in experiments with it, together with foreign data, will provide a sufficient basis to create at the beginning of the 1980s the first "demonstration tokamak reactor"— that is, a machine in which the energy output will almost match the input. He speaks of this as the Tokamak-20, which will apparently be a hybrid reactor (according to a story in the *New York Times*, 23 March 1976). A hybrid reactor uses the supply of neutrons from the plasma for producing plutonium rather than, or along with, creating heat. The current Soviet literature shows a strong interest in hybrid reactors. An article in *Atomnaia energiia* (1975:12, pp. 379–386), surveys the possibilities, and two of the major researchers in the field at the Kurchatov Institute describe this as a very attractive option (*Pravda*, 10 March 1976, p. 3). Many of the discussions of breeder reactors do envisage the possibility of specialized plutonium producers (i.e., facilities that do not produce energy), which may be needed to produce the fuel stocks for breeders if the latter become available early, grow fast, and breed slow (*Atomnaia energiia*, 1976:11, pp. 57–58). It is not explained what kind of reactors these will be, but perhaps this is one of the roles forecast for fusion. In those scenarios, however, the specialized plutonium-producing reactors play a role of limited duration.

Moreover, in the scenarios now envisaged, hybrid tokamaks would come on the scene too late to serve in this role. But one study shows that, even when breeders are widely employed, "the production of additional plutonium in hybrid fusion reactors will still be useful," a conclusion based on an assumed shadow price for plutonium, which would add enough value to a hybrid's output to justify its capital cost (*Atomnaia energiia*, 1976:11). The Electric Power Research Institute (the U.S. electric utility R and D institute) has reached an agreement with the Kurchatov Institute to explore hybrid reactor concepts. "A joint working group will plan testing of U.S. built modules in the T-20 which is now being designed in Russia" (*Nuclear News*, 1977: August, p. 144). The source goes on to say that the T-20 is the only magnetic confinement device, now existing or planned, to test the hybrid concept. The Soviet fusion program also includes intensive research on other lines as well, particularly the laser and electron-beam excitation of fuel pellets (*New York Times*, 15 January 1976). In fact, Soviet researchers are said to have made some significant breakthroughs in this approach. In the summer of 1976, L. I. Rudakov, who heads this work at the Kurchatov Institute, gave talks at several U.S. organizations in which he disclosed what are described as fundamental new ideas that might solve the problem of electron-beam fusion. These ideas are still classified in the United States, and the ironic outcome of Rudakov's presentation was an effort by U.S. authorities to control the dissemination of concepts that had been openly presented by a Soviet specialist! (*Science*, 1976:October, p. 166.)

Regrettably, it seems impossible for me to draw any conclusions about R and D decision making as reflected in the evolution of the fusion power program, or as to how fusion power is related to Soviet nuclear energy policy. This is a research area at the far frontier of knowledge. As L. M. Artsimovich, the physicist who pioneered this work in the USSR once said, "it is still not known from which branch this golden apple will fall." Fundamental breakthroughs may radically alter the priorities and directions of work at any time. Also, as an area with military implications, we can be sure that it is not discussed completely openly. An initiate, an active researcher in the field, might be able to deduce from the technical literature and from such meager bits of evidence as are available, how the Soviet managers of this effort are evaluating its potential and making their decisions about how to attack it, but the author is not in a position to do that. The remarkable thing is that, in the standard Soviet discussions of nuclear power (which, as I have indicated, have in recent years been quite straightforward and explicit), there is virtually no discussion of how the fusion effort is managed or of how it is seen as fitting into nuclear R and D policy in general.

NUCLEAR SAFETY

Special comment is in order as to the influence of environmental and safety considerations on Soviet nuclear R and D effort. It is a commonplace in Western analyses of the Soviet nuclear program to assert that safety and environmental considerations weigh much less heavily on Soviet decision making than on ours. This position is no doubt basically correct; it does seem that Soviet nuclear power policymakers have not really faced up to most of the hard questions that are so important in Western concerns about nuclear safety. For example, little credence is given to the possibility of a loss-of-coolant accident. As one Soviet source says,

> Generalization of the operating experience of the uranium-graphite channel type reactors in the USSR shows that unexpected, instantaneous rupture of large-diameter pipes and drums, designed according to accepted standards, and manufactured according to the technology of high pressure vessels, at moderate pressures and temperatures under a system of verification and inspection, are extremely unlikely. [*Atomnaia energiia*, 1971:October, p. 239]

The conclusion of a U.S. delegation in 1970 was that the basic philosophy on safety was to "omit intricate design backups for accidents determined to be virtually impossible. In this category are propagation of coolant flow loss beyond a single fuel assembly and progression of small accidents into large ones with bulk fuel meltdown and redistri-

bution and explosive energy release. . . . In many cases, design requirements are limited to likely events or malfunctions and unlikely events
are not designed for. . . . In summary, Soviet safety practice is focussed
on reliability and operability of components and systems through adequate design margins and engineered safeguard features which prevent
small accidents from occurring, and for cases considered to be reasonable, from developing into serious accidents" (AEC, 1970, pp. 16–18).
There is certainly no special sensitivity about the increased risk associated with locating nuclear plants near heavy concentrations of population. A. Grigoriants, head of Minenergo's nuclear power administration seems to accept that it would be uneconomic to locate nuclear
TETs 30–40 KM from cities (as is now done for conventional nuclear
plants) and sees them as being located on the edges of cities, though
he makes the concession that they should have containment shells
(*Pravda*, 27 March 1977). This is not to say that Soviet officials have
no concern with safety; there is considerable Soviet literature on safety
as an aspect of nuclear power plant design. The twentieth anniversary
volume on atomic power (*Atomnoi energetike XX let*, 1974) includes
a chapter on problems of safety. There are reasonably frequent articles
in *Atomnaia energiia* on various aspects of safety and environmental
effects. A recent book (which unfortunately I have not yet been able
to obtain) covers the whole area (Emelianov, 1975). But the focus in
most of this literature is on safety as an aspect of the normal operation
of plants. The reviewer of the Emelianov book says that it does provide techniques for evaluating the risks associated with various failures, but whether the Russians use anything like the fault tree design
used in U.S. studies is not clear.

NUCLEAR REPROCESSING

It is difficult to find direct evidence on the Soviet attitude toward
the plutonium problem that is an important element in Western fears
about the breeder reactor. The Soviet literature does not even mention
the danger of diversion or terrorism as a potential difficulty domestically, but the Soviet authorities *are* concerned with the dissemination
of plutonium as an aspect of national security. Perhaps the best clue
to their attitude is how fuel for the reactors supplied to the Eastern
European countries and to Finland has been handled. It is apparently
required that spent fuel be sent back to the USSR for reprocessing.
An article in *Atomnaia energiia* (1975:July) describes the container

and transport systems used for shipping spent fuel and speaks of its being shipped to the USSR, even mentioning the problem of shifting the gauge of railroad cars at the Soviet border. But since R and D on reprocessing is conducted within the CEMA framework under a Council on Reprocessing of Spent Fuel of the Permanent Commission on Peaceful Uses of Atomic Energy, the USSR must not be ruling out entirely the diffusion of reprocessing technology to other Communist countries. Finland also returns spent fuel to the USSR for reprocessing—a feature of the contract with which the Finns are pleased (*Nuclear News*, October 1977, p. 72), since that relieves them of the problem of storing or processing the wastes. There is, however, a curious article in *Atomnaia energiia* in 1976, in which two Finnish researchers analyze the possibility of loading a VVER reactor with plutonium (Kaikkonen, 1976). If the Soviet Union had a clear policy on plutonium and were adamant about preventing its spread, it seems doubtful the publication of such an article would be permitted. The USSR is reported to have offered to sell uranium fuel to the Japanese (*New York Times*, 11 November 1977). It is possible that the story misinterprets the situation and that all that is being offered is toll enrichment. But one can understand that if uranium itself is being offered the Japanese would probably be pleased, like the Finns, to be able to send spent fuel back to the supplier. Despite past Soviet practice in not supplying domestic uranium to other countries (except in the case of Finland) there is a possible rationale for doing so. Soviet nuclear plans are counting heavily on a considerable supply of plutonium, and from that perspective it might be better to have uranium converted to plutonium in other people's reactors rather than sitting at home.

The Czech A-1 reactor, described earlier, produces plutonium in significant quantities. I have seen no statement as to whether the spent fuel from that reactor must be returned to the USSR, nor do we know whether the fuel (in the form of unenriched metallic uranium) is prepared in Czechoslovakia (and hence owned by the Czechs) or is produced in the USSR from Czech uranium ore sold to the USSR. It is said that the new VVER stations being built in Czechoslovakia will use Czech raw material which will be processed "komplektno" (i.e., probably refined, enriched, and fabricated) in the USSR (Neumann, 1972). That statement might be interpreted as implying Czech ownership and toll enrichment, but does not have to mean that. An account by the former General Director of the Czech uranium mines (who left Czechoslovakia in 1970) says that the USSR has been adamant in refusing the Eastern Europeans the right to produce their own nuclear fuel (Bocek, 1974).

Finally, the anniversary survey (*Atomnoi energetike XX let*, 1974, p. 199) suggests that once the breeder is developed, such reactors will be used in the Eastern European countries along with the VVER. Moreover, the Soviet planners expect the Eastern Europeans to share in this technological development. As one source puts it "the most important goal of cooperative efforts to develop atomic power engineering should be the creation of breeder reactors" (*Voprosy ekonomiki*, 1976:6, pp. 70–79).

The most recent line is that the USSR supports the idea of regional reprocessing centers for spent fuel. In a report reacting to the U.S. efforts at the Salzburg conference in 1977 to win support for slowing down plutonium technology, A. M. Petrosiants supports such centers and says "we can cite as an example the region of member countries of CEMA where the majority of the problems of organizing the external fuel cycle have long been decided jointly with the participation of all interested countries" (*Atomnaia energiia*, August 1977, p. 85). Despite the hazards of being the depot for spent fuel, one can imagine that the USSR would accept this role both as a counter to spread of nuclear material and as a way of accumulating plutonium for fueling the breeder when it is ready. And it is surely in a strong enough position that "joint decisions with the participation of all interested countries" are unlikely to violate Soviet wishes.

It is not at all clear where the USSR now stands in the development of reprocessing technology. There must be some experience with processing wastes to recover plutonium. The "Siberian" reactor at Troitsk has apparently always been under the control of the military, and since it is an early version of the graphite moderated reactor, its role must have been, at least in part, plutonium production for military purposes. This implies some experience with reprocessing, though this technology would be unlikely to meet requirements for a commercially acceptable technology. One recent statement intended for external consumption is that "the search for reliable means of getting rid of the waste from atomic power stations is continuing in the USSR. Such waste is now being buried at great depths in limestone caves which are geologically sealed off and where there is no risk of water being contaminated" (*Soviet News*, 3 May 1977, p. 4). The answer as to why research work on reprocessing is being conducted jointly with the

Comecon countries may be that the USSR needs the scientific and technological assistance of these countries. As one source says: "analysis shows the need for substantial cooperation in the process of integrating the external fuel cycle" (*Vestnik ANSSSR*, 1976:11, p. 60). It is really striking to see how little is said about nuclear processing technology, and it may be that this essential aspect of the overall breeder strategy has not been given the attention it deserves.

To summarize, it is probably true that the nuclear technologists have so far been able to make decisions with relatively little concern over safety aspects. There is no doubt some skepticism and opposition to nuclear power from various places in the bureaucracy, especially from people with vested interests in other energy forms and other R and D projects. The distinctive feature of the Soviet setting is that there is no public initiative or public forum that would raise safety and environmental objections or question the arguments and design approaches the nuclear power people offer in answer to safety concerns. The usual response from these people is that the dangers to health and personal safety from nuclear power are less than those from alternative energy sources, and this is a point not difficult to make when the argument is limited to wastes and emissions from routine plant operations. But the attitude is probably changing, and safety considerations are likely to come to have a more inhibiting effect in the future. It may be that an increasingly intimate involvement with Western specialists has heightened Soviet awareness of safety problems. The Finns insisted on fitting their VVER-440 reactor at the Loviisa plant with a containment structure, even though the layout of the Soviet reactor meant that the containment structure had to be much larger than for a Western reactor of equivalent capacity (*Nuclear News*, October 1977, p. 73). That kind of insistence must in the end be persuasive. The latest addition to the Novo-Voronezh plant (a 1,000 MW VVER reactor) will, unlike all previous VVER reactors, have a containment structure (ERDA, 1974, p. 7). In a recent statement describing the future growth of nuclear power, P. S. Neporozhnyi says that it is planned to increase safety measures, which will raise capital costs and thus make it imperative to achieve higher utilization (6,000 hrs/yr) to maintain the competitive standing of nuclear power (*Elektricheskie stantsii*, 1977:8, p. 3).

A recent Soviet article on safety concludes with a paragraph that I doubt could have appeared in earlier years. The reviewer endorses the position that nuclear plants are safe in operation but concludes by quoting approvingly a Western writer who says that the power plant itself is but the tip of an iceberg. He asserts that of the three main systems—the power plant, the external fuel cycle, and the handling of wastes—Soviet analysts have considered only the first. His final word is that "ecological analysis of the whole fuel cycle of nuclear power is the most important task of the upcoming years" (*Atomnaia energiia*, 1976: October, pp. 235–238). But so far this shift in attitude does not seem to have gone far enough to have any impact on three central features of the Soviet program. Current efforts are still solidly based on faith in the breeder. There is a serious development program and concrete plans for utilization of by-product heat and heat from specially designed reactors for both space heating and industrial processing. Nobody seems to find anything objectionable in the location of large plants near population concentrations.

CONCLUSIONS

Looking back over this review of nuclear development, what generalizations might be made? First, the nuclear program demonstrates unequivocally the Soviet capability to develop a new technology at the very forefront of scientific advance and bring it to successful commercialization. They have a working technology, and it seems to have been adapted to the specific Soviet economic conditions and goals. In general, Western visitors are impressed with what the Russians have accomplished. This comes through clearly in the two reports on Soviet power reactors produced by U.S. delegations sponsored by the AEC and ERDA (AEC, 1970, and ERDA, 1974), already cited. Moreover, the Soviet program is not merely derivative. I believe it is possible that the VVER reactors owe something to the inspiration of Western choices, but the RBMK is claimed by the Soviet authors, and acknowledged by Westerners, to represent a distinctive line of development and an original achievement. When the RBMK first came into use, Academician Dollezhal' claimed that "no other country has solved this problem" (*Atomnaia energiia*, 1976:2, p. 117), and Westerners do not dispute this position. A story in *The Financial Times* (11 July 1975) says that British nuclear engineers who had toured the USSR nuclear installations were very impressed with the RBMK reactor and found a number of ideas and innovations that would be helpful in their own program. As indicated earlier, the USSR is one of the world leaders in breeder development. There has been a substantial interaction with the national programs of other countries, so that the Soviet breeder program does not represent a really independent direction. But the USSR appears to have come about as far as Great Britain or France in solving technical problems and embodying the technology in workable designs.

If there are doubts about the technical achievement, they are associated as usual with the more downstream stages of producing and mastering the technology in use. There is some indication that each of the plants so far built has had some element of being an experimental installation, in which problems not taken care of in the design or manufacturing process have to be taken care of in the construction of the plant. We also see indications of the failure to get the technology developed as an integrated process. For example, the turbines for these plants have so far been mostly standard or slightly modified designs, not really optimized or integrated into the station design. The Khar'-kov turbine factory produced turbines for the various early stations on a more or less ad hoc basis making heavy use of subassemblies from standard series models (*Elektricheskie stantsii*, 1971:6, pp. 2–7). The BN-600 will use three standard K-200-130 turbines already widely used in conventional stations (ibid., pp. 118–119). I suppose that must be because it will have steam at much higher temperature and pressure.

A Western authority comments that it seems a sign of technical backwardness to equip a 440 MW reactor, as they do, with two turbines. U.S. reactors have for some time been connected to single turbines of 800 MW and more (Sporn, 1968, p. 35). The Leningrad station has two 500 MW turbines and generators for each 1,000 MW reactor. Only with the move to the 1,000 MW VVER will special turbines matched to the capacity of the reactor be designed. This turbine will have 1,500 RPM speed and at the beginning of the 10th Five Year Plan was in the process of being produced (Nekrasov and Pervukhin, 1977, p. 118).

As another indication that this technology has not been satisfactorily assimilated and mastered at the industrial level, the USSR is very interested in getting foreign assistance to build the new nuclear plants. At one point the West Germans were invited to consider building nuclear plants on Soviet territory with some of the power output going to Germany. This may only indicate a desire to export energy without letting uranium leave the country, but it seems more likely that it reflects great difficulties in building nuclear plants themselves. The USSR has asked the Japanese to supply control equipment and on one occasion sought to buy reactors built to Soviet specifications (*Nuclear News*, March 1976, p. 60). Czechoslovakia seems to have won for itself a significant role, not only in supplying for the VVER stations to be built in Eastern Europe equipment that has already been mastered, but also in developing the equipment for the succeeding generation of pressurized-water reactors, including the circulation pumps for the VVER-1,000. It also supplied items for the RBMK-1,000 (Neumann, 1972, p. 20). The Russians have expressed great interest in importing British nuclear equipment. The USSR has not tried to export its nuclear power technology, except to a captive market in Eastern Europe and to Finland. Offers have also been made to India, Libya, and the Philippines. It is thus impossible to appeal to the test of competition as a way of judging whether Soviet nuclear technology is on the same technical and commercial level as that achieved elsewhere, though my guess would be that it is not.

Nuclear R and D underlines some characteristic features of Soviet R and D already noted and suggests some clues as to what determines the degree of success of an innovation. As with other examples examined in earlier chapters, the Soviet approach in R and D is to move fairly early into expensive experimental facilities and to use them as a test bed to develop the technology. The experience of the Czech A-1 reactor seems to reflect this tendency to premature commitments. One American familiar with all the national breeder programs says that the Soviet Union "prefers to build the reactor first, then see if it can be made to work," in contrast to the U.S. approach of testing many design variants and actual components (*Science*, 30 January 1976, p. 371). A British observer notes that construction of the BN-600 was started well before many of the design decisions had been made. An inclination to make expensive and far-reaching expenditures well in advance of clear-cut answers to technical uncertainties continues right on through all stages of the R and D process. Although the BN-600 is not yet in operation, the long lead times involved mean that work on the next generation, the commercial version, has already begun. The breeder will be economically competitive only if its size can be increased, and the present development work in the USSR envisages a very large upward jump to a 1,600 MW reactor for the commercial version. The French and British reactors of that generation will be 1,200 and 1,300 MW respectively (*Science*, 26 December 1975, p. 1281). One author claims that it is urgently necessary to begin actual widespread construction of such units in the immediate future. He acknowledges that more work must be done to make the steam generator reliable and that it needs more development work before it can be built to the scale required in a 1,600 MW reactor, but "we have no doubts that the difficulties in this area can be overcome" (*Pravda*, 27 March 1977, p. 2). It is the steam generator, of course, that presents the most troublesome uncertainty in all breeder programs.

One of the important features of all the R and D cases involving electric power, which tend to be more successful than what we have found for the coal industry, appears to be the power of Minenergo as the one big buyer. Minenergo also seems to have a rather intimate involvement in the R and D decision making and in the actual development of the technology. The ability to exert such influence seems to depend on how the responsibility passes between the R and D hierarchy and the client hierarchy as a program advances. Little is known about this in general, but some interesting features can be noted in the relationship between Minenergo and the organizations that have developed some of the new energy innovations. Minenergo seems to have had a close involvement with the nuclear program from the beginning. In 1951 its project-making organization, Teploenergoproekt, participated in designing the Obninsk station, the original test facility. It also did the design work for the first block at Novo-Voronezh (the first VVER) and the first block of the Beloiarsk station—both of which were still prototypes. In 1966 it became the general designer of all nuclear power stations (*Teploenergetika*, 1974:4, p. 5). In contrast to the cases of oil or gas or coal equipment, the R and D organizations in nuclear power were incapable of designing a test facility on their own, and a certain symbiosis between the developers and the client grew out of this dependence. It is said that in the development of the breeder, developmental responsibility passed to Minenergo with the decision to build the BN-600. The BN-350 was developed by the State Committee for the Peaceful Uses of Atomic Energy. The BN-600 is still one step away from the form of the plant expected to be put into series production, and it is said that Minenergo is giving great attention to designing it to conform to its own concept of what is necessary in a commercial plant (Rippon, 1975, p. 572).

I suspect that Minenergo also had a strong influence on the development of the RBMK-1,000. It is a development from the Beloiarsk plant, which, according to the source just cited, is under the control of Minenergo. Also we know that the RBMK-1,000 was designed by Gidroproekt (the biggest project-making organization in Minenergo).

When the existence of the RBMK in more or less fully developed form was disclosed, it came as a great surprise to American observers. An American delegation from AEC toured the Soviet power stations and test facilities in 1970 under the auspices of the State Committee for the Peaceful Utilization of Atomic Energy and were given no hints that this type of reactor was being developed. It thus came as a great surprise in 1971 when the Leningrad station was announced, not as a proposal but as a mature technology, embodied in a 1,000 MW reactor. Perhaps the reason the State Committee did not mention the RBMK at the time of the 1970 visit is that it was the model being developed by the competition.

We also know that Minenergo project-making organizations have played an important role in the forecasting studies that guide nuclear strategy. Academicians Dollezhal' and Melent'ev say that their forecasts are based on work done at Energoset'proekt (*Vestnik ANSSSR,* 1976:11, p. 51).

It is claimed that intimate cooperation has been the rule in the development of nuclear power. At the 25th Party Congress, Aleksandrov cited the nuclear power case as an example that could show the way to such cooperation elsewhere in the economy, implying of course that such cooperation does *not* exist in other areas:

> Branch institutes or plants with their own personnel should be included in the institutes' projects at an early stage, and a significant part of the work—from initial exploration through the final application of the research—should be done jointly. The experience in the development of the nuclear energy industry shows that with this sort of work the complexities of industrial application disappear. Moreover, industry immediately obtains its own well-trained personnel who apply the innovations in practice. [*Izvestiia,* 27 February 1975]

The most distinctive peculiarity of the Soviet nuclear R and D program is probably the more casual position taken on nuclear safety. This has the effect of permitting numerous lines of research and development that are blocked in the United States. There is here a difference of policy and opinion with potentially very important consequences for the future. If the Soviet leaders are correct that the environmental and safety hazards are far less serious than many in the United States consider them to be, the USSR may well be in a position to exploit the potential of nuclear energy much more profitably than we can and to move well ahead of the United States in depth and variety of experience with this technology. I do not think there is any way to judge which view regarding safety is correct, since we are talking about the uncertainties inherent in a new technology. It is a bit worrisome, however, to see how little thought appears to have been given in the USSR, even today, to alternative long-range strategies and to the relationship of fission power with fusion. The commitment to the breeder has the distinct flavor of a technological fix, decided on early in the game and transformed into dogma. It is also disturbing to see how little public discussion there is on any of these issues; these choices are announced ex cathedra as foregone conclusions.

I am not sure that we should worry much about the consequences if the Soviet view turns out to be right. If the USSR demonstrates the safety and economy of the breeder and accumulates long experience of safe operation with reactors that are less carefully engineered for safety than ours, then the United States can benefit from that demonstration and ought to be able to catch up very quickly, adapting Soviet experience. We might think of Soviet nuclear power policy as a kind of experiment inflicted on the Russian people that we would not choose to risk ourselves, but from which we can greatly benefit if the experiment is a success. And as long as the United States maintains an exploratory program at the frontier of breeder technology, we will have the base and industrial flexibility to move decisively ahead of the USSR when it is decided that the breeder is safe. When they have accumulated enough reactor hours so that it is possible to make better evaluations of hazards in reactors, we can make better-grounded decisions about how to design around these hazards. And in breeder technology it is not a U.S.-Soviet race that is taking place; the British and French are moving at more or less the same pace as the USSR, so that we need fear no commanding Soviet lead or Soviet monopoly in that area.

On the other hand, if the Russians are wrong and the dangers of their choices are demonstrated by some spectacular accidents or failures, we can congratulate ourselves on having avoided that mistake. Unfortunately, however, the correct interpretation of any such disaster in the USSR would probably be that it reflects mistaken design philosophy and engineering rather than inherent infeasibility. But that may be a hard case to make convincingly and any Soviet failure could choke off what might ultimately be seen to be the right way to go. Certainly the kind of failures I have in mind (such as a spectacular accident in a nuclear TETs on the edge of a major city) could not be concealed, and it would no doubt be very difficult for the advocates of nuclear power to quiet the internal opposition that such a failure would raise.

Regarding the matter of nuclear proliferation, there is probably less danger that the Soviet nuclear policymakers could make a major miscalculation that would redound to our misfortune. Except for the ambiguity in their handling of the reprocessing issue and their willingness to create large stocks of plutonium, they seem to be much more cautious on this score than most Western countries. In their external pronouncements, Soviet spokesmen are taking a hard line against the export of nuclear technology. They hang many extraneous issues on the point (such as the illegitimacy of commercial motives), but assert that "the widespread development of nuclear power conceals within itself the potential danger of nuclear arms" and that there should be strict limits on the export of the most dangerous element—i.e., the reprocessing technology (*Pravda*, 8 July 1977). They support the idea of regional processing centers and fuel banks. They emphasize use of the controls of the IAEA and indeed seem to have more faith in those controls than some Western observers do. And in the private efforts to control proliferation through the committee of exporters, they seem to be on the side of caution and restraints on exports.

6 Technologies at Early Stages of Development

The purpose of this chapter is to look at several technologies that are still in fairly early stages of development and commercialization. Some of them—coal conversion, slurry pipelines for coal transport, and power transmission at ultra-high voltages—are closely related to one of the most important current goals of Soviet fuel policy, namely drawing eastern coal into the national energy balance. Another—MHD—is also closely related to important current concerns—i.e., it is the major approach to obtaining further reduction in the heat rate in electric power generation. These technologies are thus potentially quite significant in quantitative terms; how successful the Russians are in getting them commercialized will have a significant impact on the shape of energy policy in the next decade. Three other technologies to be discussed in this chapter—tidal power, geothermal and solar energy—are much less important for overall fuel policy, and their commercialization is more remote and less urgent, but they may ultimately make some palpable contribution to energy supply and be important innovations in specific cases. These seven examples are all interesting case material for illuminating Soviet choices about allocating effort along the spectrum of different development stages and in moving along this spectrum from one R and D stage to another.

SLURRY PIPELINES

The USSR is faced with some very large coal transport needs that cannot be met by the existing railroad network. In particular, as oil becomes scarcer and gas more expensive, current thought on fuel policy favors a larger role for the major coal basins of the East—Ekibastuz, Kuzbass, and Kansk-Achinsk. Most of this coal would have to be used in the West, so that if coal is to substitute for oil and gas, some way must be found to transport it over long distances; the existing rail links cannot possibly take it on. In the USSR, as in the United States, the relative advantage of rail transport and coal slurry pipelines for long-distance coal transport is a controversial issue, but it is thought that coal slurry lines may well play an important role in transporting

170

Siberian coal. As one author says "in the conditions of the Soviet Union, hydraulic transport of coal will be developed above all in the Kansk-Achinsk deposit. Hydrotransport may also be used for shipment of coal to power stations located in Siberia and other regions of the country, as for example the Ural region" (Ushakov, 1972, p. 172).

There are two related questions—how successful the USSR will be in developing slurry pipeline technology and how competitive slurry pipelines, once mastered, will be compared to railroads. In the United States, the technology of slurry lines seems to be thoroughly in hand, and the major determinants of their application involve economic and institutional considerations. At issue are whether they should be organized as common carriers, whether they should have the right of eminent domain, whether the government should protect railroads against competition from them, and whether in any given situation their somewhat distinctive requirements, such as an adequate water supply, make them noncompetitive. The general conclusion in the United States regarding their competitiveness seems to be as follows:

> The two low-cost methods [for transporting coal overland] are slurry pipeline and rail. The overlap of .3 to .7 cents per ton-mile for slurry and .4 to .9 cents for rail is realistic. Neither is superior to the other in any broad spectrum. Yet, for specific cases one will undoubtedly be preferable to the other, though likely by narrow margins. [U.S. Bureau of Mines, 1975, pp. 23–24]

The first large-scale slurry line in the United States, the Cadiz-Eastlake line in Ohio, which was considered a decided technical success, ceased operation because a cut in railroad tariffs permitted the railroads to win back the traffic. In the USSR, it seems, most of the institutional issues important in U.S. debates are absent, but there are still important technological questions to be settled.

Several hydraulic coal transport systems have been in operation for some time in the USSR. Most involve very short hauls and only move coal from a mine to its cleaning plant. One which began operation in 1967 sends coal from the Inskoe mine in the Kuzbass to the Belovo power station in West Siberia, and another (operating since 1966) sends coal from the Iubileinaia mine to the West Siberian metallurgical plant. Each is 10.5 km long; the first has a capacity of 1.2 MT per year, the second 3.9 MT per year (TsNIEIUgol', 1976, p. 4). In both these cases the coal is mined hydraulically, so that transport to the user is an extension of the transport approach used within the mine (Smoldyrev, 1970, p. 1). The Iubileinaia installation uses five 350 mm (approximately 12-in.) pipelines, of which three work and two are in reserve, implying that the line must be unreliable (Smoldyrev, 1967, p. 71). Other sources also speak of the rapid erosion of the pipe. These are not slurry lines, properly speaking, since they operate with very coarse coal (sizes up to 50 mm) and a low concentration (15 percent or less) of coal in the mixture (TsNIEIUgol', 1976, p. 4).

Some more ambitious applications have also been proposed. A project for supplying the Belovo station in West Siberia with an additional 360 thousand tons of coal from a mine 25 km away has been under consideration since at least the mid-sixties. A still more ambitious project envisages a 420 km pipeline from the Lugansk mines to the PriDnepr regional power station in the Ukraine. That project would use 400–500 mm pipe and have a capacity of 4–5 million tons per year (*Energetika i transport*, 1976:6, p. 7; and Smoldyrev, 1967, pp. 82–83). It would thus be a fairly close analogue to the Black Mesa line in the United States, which transports 5 million tons of coal a year in a 287 mile line using 18-inch pipe. Other projects mentioned in the literature are lines to the power stations at Dobrotvorskaia (61 km), Mironovskaia (88 km), Staro-Beshevskaia (64 km), and Kurakovskaia (80 km) (Smoldyrev, 1967, p. 83; and TsNIEIUgol', 1976, p. 5). These projects all envisage moving coal in slurry form—i.e., the particle size for various projects is in the 0–2 mm, 0–3 mm, 0–6 mm classes, and the coal concentrations would be 40–50 percent. Whether any of these projects have actually been completed is unclear, though my guess would be that they have not. A 1973 statement says that 300 kilometers of coal transport pipeline are actually being operated (*Ugol'*, 1973:3, pp. 37–38), and a 1978 source says that 10 coal pipelines are in operation (*Truboprovodnyi transport*, vol. 7, 1978, p. 89). But what would seem to be an authoritative recent statement by a Gosplan author says explicitly that no high-concentration lines have yet been created and mentions only the two Kuzbass lines mentioned earlier as actually being in operation (Iu. Bokserman in *Planovoe khoziaistvo*, 1977:10, p. 100).

The use of high-volume, long-distance slurry lines for moving eastern coal to the Ural or to the European USSR would thus be a large step upward from any previous experience. Economic studies by the research organizations comparing various transport modes for moving eastern energy westward make slurry pipelines look very attractive. Numerous proposals have been considered—based on Kuzbass, Ekibastuz, and Kansk-Achinsk coal, using pipe diameters of 1220, 1420, and 1620 mm, with capacities from 50 MT to 100 MT per year. Slurry pipelines are appealing in terms of labor inputs, metal requirements, energy efficiency, and overall transport cost. Most of the evaluations conclude that pipeline transport is likely to have a small cost advantage over rail or electric power transport (*Planovoe khoziaistvo*, 1977: 10, p. 101; TsNIEIUgol', 1976, pp. 13–19; Ushakov, 1972, p. 174), and some studies show large cost savings.

But all available evidence suggests that the USSR is not very far along in creating slurry pipeline technology. As indicated there were no large, high-concentration slurry lines in operation as of the mid-seventies. The main reason for this seems to be that Soviet industry has not yet produced the proper pumping equipment. One of the early projects apparently envisaged the purchase and use of the same pumps that were used on the Cadiz-Eastlake line (Smoldyrev, 1967, p. 82); a later commentator says explicitly that Soviet-produced pumps capable of moving slurries "operate at low pressure and have design defects that do not permit using them in long-distance systems with high concentrations of solids" (*Planovoe khoziaistvo*, 1977:10, p. 101).

In this situation it seems indispensable to get on with a demonstration project to test equipment and design ideas, but the Soviet planners seem to have been very cautious in moving to any demonstration effort. One author suggested in a 1972 book that the appropriate step would be to design a large-diameter experimental line (1420 mm) from some Kansk-Achinsk mine to a nearby large power station as a first step in developing the data needed for projecting a line from Kansk-Achinsk to the Ural (Ushakov, 1972, p. 175). But that idea has not been followed up, and the proposals now actively being considered involve lines of smaller diameter but greater length. One article mentions two such projects—one from Leninsk-Kuznetskii in the Kuzbass to Barnaul would move 5–6 MT of coal per year, and one from Borodino to Krasnoiarsk (150 km) would move 3–5 MT of Kansk-Achinsk coal per year (*Planovoe khoziaistvo*, 1977:10, pp. 101–102). Another project mentioned more recently calls for a line from the Kuzbass to Novosibirsk 250 km long, using pipe 429 mm in diameter to handle 4.3 MT per year (*Planovoe khoziaistvo*, 1978:11, pp. 21–23). It is said that the design for this line has been completed and that experiments with equipment are being carried on at the Ramenskoe test facility of VNIIPITransprogress (*Stroitel'stvo truboprovodov*, 1978:11, p. 37). Fuel policymakers are apparently serious about this kind of line, and a group of experts headed by A. Lalaiants, the deputy chairman of Gosplan in charge of energy policy, has recently been in the United States, especially to learn more about experience with the Black Mesa line.

Lines with those capacities and using relatively small diameter pipe will do little to settle the issues involved in designing a line with the 50–100 MT capacity needed to send eastern coal to the Ural region or to the Center. For dealing with that problem, all that was envisaged for the 1977–1980 period was research and experimental work leading to the establishment of specifications for pumping equipment and fittings and designing the technical schemes for such a line. A technical-economic analysis for a 25 MT/year line 2800 km long is said to have been completed, as well as preliminary estimates for an 85 MT line from the Kuzbass to the European USSR (*Planovoe khoziaistvo*, 1978: 11, pp. 21–23).

The idea of an experimental slurry line for the short haul from the mine to the Berezovo power plant (the first of the big power stations to use Kansk-Achinsk coal) has apparently been rejected in favor of a container pipeline. Some experiments in the early seventies convinced Soviet transportation planners that container pipelines have a big future, and a 1974 decree of the Council of Ministers called for 28 experimental (*opytno-promyshlennyi*) container-line installations to be built, including some in the coal industry (*Current Digest of the Soviet Press*, vol. XXVI, No. 32, pp. 12–13). Container pipelines do not really seem suitable for coal transport since even a 1400 mm container pipeline would move only 5–10 million tons a year (ibid.). This seems very small in relation to the annual fuel consumption of one of the 6.4 GW stations intended for the Kansk-Achinsk basin (something more like 25 MT) and very extravagant of metal compared to a much smaller diameter slurry line of equal capacity. But for some reason or other the proponents of container pipelines have been extremely successful in convincing the top leadership of their potential and have succeeded in getting a high priority for development and demonstration of this technology.

As a case of R and D, slurry pipelines reveal a slow and cautious attitude and relatively little sense of direction on the part of the program managers. One of the most interesting features of the literature is the many contradictions and gaps in the accounts given by different authors. One source claims that actual design (*proektirovanie*) was completed for a large-capacity line from Kansk-Achinsk to the European USSR (*Truboprovodnyi transport*, vol. 6, 1976, p. 45), but later discussions seem completely oblivious to this. It is possible that the slowness and uncertainty in moving ahead reflect not dilatoriness but uncertainty as to just when and in what quantities Kansk-Achinsk coal will be used in the West, since the questions of how it will be burned and whether it can be converted into a transportable form have not yet been settled. I have found no comment as to whether Kansk-Achinsk coal would be sent via pipeline without prior conversion. It seems that pipeline transport *would* permit the sidestepping of prior conversion and that the high water ballast that makes it uneconomic to ship by rail would be less of an obstacle in a slurry pipeline.

The strongest impression one has from all this discussion is of how little R and D work has been done and how vague and irresolute policy seems to be in setting any kind of directions and goals for moving ahead with some kind of demonstration work on slurry pipelines, given the urgency of finding a solution to the problem of transporting Eastern coal.

CONVERSION OF SOLID FUEL

Conversion of solid fuel constitutes a very broad area of technology that includes such processes as coking, gasification, liquification, solvent refining, and many others. It is not a frontier technology in the same sense as some of the others discussed in this chapter, since various versions of many of these processes have been in commercial use for a long time—e.g., gasification and hydrogenation. On the other hand it is far from being an established technology, because the current situation creates new demands with respect to scale of operation and type of product and raw material, which cannot be satisfied by the established technological base. This is the case in the USSR as well as in the Western countries. The USSR has had a long history of experimentation and research in solid fuel conversion, sometimes with commercial application—it has carried on commercial processing of shale and has produced synthetic fuel and gas from coal on a significant scale. Today, however, Soviet energy policy faces quite a new problem: the transformation of the cheap strip-mined lignites of the East, notably those of the Kansk-Achinsk basin, into fuel forms that can be transported. Kansk-Achinsk coal is not transportable in its present form at all, and simple forms of treatment (such as an emulsion-coating to prevent drying and spontaneous combustion) do not solve the problem of the low heating value per unit weight, which makes it very costly to ship over long distances. For this coal to assume its anticipated place in the fuel balance, it needs to be converted to some form of transportable fuel with a higher heat content—liquid, gaseous, or solid. Given the relative abundance of natural gas resources in the USSR, the most useful products of solid-fuel conversion in the Soviet economy will be solid and liquid fuel (in contrast to the United States, where one of the main goals for R and D on synthetic fuels is to develop a source of gas). A review of the Soviet experience on solid-fuel conversion is illuminating, both for historical perspective on the R and D process and as an illustration of current R and D performance and its interaction with energy policy.

The USSR has a fairly long experience with coal conversion along two main lines—gasification and what the Russians call "energy-technological processing."

Gasification

The original Soviet rationale for gasification of coal was to create a gas supply at a time when Soviet planners were still unaware of the large resources of natural gas they could tap.* The main objectives were underground gasification as an alternative to mining coal, synthetic production of gas for household use, and gasification of high-sulfur coal to provide a fuel for gas turbines. (There has also been a long-standing interest in gasification of shale, but that will not be discussed here.) By authoritative Soviet accounts, the underground gasification efforts have proven a failure. According to Grafov, Deputy Minister of the Coal Industry, "for decades we worked on this—institutes, groups of specialists, engineers. . . . Unfortunately, during all those years we obtained no positive results. The gasification process turns out not to be controllable, and the heat value of the gas obtained is low" (Kirillin, 1974b, p. 46). An informative description and evaluation of Soviet underground gasification efforts—somewhat more positive than given here—is available in Gregg (1976). The major effort to produce synthetic gas from coal for household use was a high-pressure gasifier plant at Shchekino, which generated gas with a heat content of about 4,500 KKal/cubic meter for pipeline shipment to Moscow. That project used Moscow basin lignite, but experimental work was also done with lignite from the Kansk-Achinsk basin.

*The history of gasification efforts is sketched fairly extensively in Altshuler (1976), and this section draws heavily on that source.

None of that work has much relevance to the present problem of processing Kansk-Achinsk coal. None of the processes involved pipe-line-quality gas, and the conversion of Kansk-Achinsk coal to low-BTU gas is uneconomic even for local use (except perhaps in a few isolated areas such as East Siberia) because of the competition of natural gas from West Siberia. Moreover, this coal has a low enough sulfur content that there is no obstacle to burning it directly. Gasification of coal to remove sulfur and provide a fuel for gas turbine use has lost its rationale, both because of the competition of natural gas and because (as shown in Chapter 3) the Soviet Union has not been very successful in developing gas turbines as prime movers for electric power generation. Thus, most Soviet R and D work on gasification performed through the mid-fifties was unsuccessful, or has been made irrelevant by changes in the energy policy environment. Since the big switch to oil and gas that began in the early sixties, R and D on solid-fuel conversion has been carried on mainly as a kind of defensive effort against future developments or to keep up with other countries. According to Z. F. Chukhanov, an eminent researcher in this field, work on processing solid fuel has had an extremely low priority for the last 15 years (*Vestnik ANSSSR*, 1976:9, p. 15).

Complex processing

Much more relevant to the current problem of utilizing the Kansk-Achinsk coal is work on what the Russians call "energy-technological processing of coal," producing both fuel and chemical raw materials. Soviet experience in this area grows basically out of earlier efforts to produce synthetic liquid fuels. A large effort was made during the Second World War to produce synthetic motor fuel by destructive distillation of lignites, and several plants were built for this purpose (Krapchin, 1976, p. 133). The motor fuel produced was both low in quality and very expensive, however, and when the discovery of large oil resources eased the problem of liquid fuel supply in the fifties, these plants were modified to use their output of tar products for chemical purposes. There are now great hopes for applying this experience to the processing of Kansk-Achinsk coal.

An experimental program of fairly long standing has been carried on by the Krzhizhanovskii Power Institute in Moscow (ENIN), the Institute for Research on Mineral Fuels (IGI), and a number of other research organizations. A pilot operation (described as an *opytno-pro-myshlennaia ustanovka*) using Moscow lignite was created at the

Kalinin power station in 1957, and an analogous unit was to be put into operation working on peat in 1958 at the Sverdlovsk power station in the Ural region (*Elektricheskie stantsii*, 1958:2, p. 5). Both were based on ENIN designs. I have seen no explicit statement as to the size of either, but a later source says that the pilot plants for energy-technological coal processing have had capacities of 100–200 tons/day or 26–55 thousand tons/year (Z. F. Chukhanov in *Vestnik ANSSR*, 1976:9, p. 118).

The next step in the commercialization of energy-technological processing of Kansk-Achinsk coal is to be a pilot plant adjoining one of the Krasnoiarsk heat and power combines (TETs-2) intended to process about one million tons of lignite a year. This plant is apparently to consist of a single unit of the ETKh-175 converter designed by Teploenergoproekt, one of the big design organizations in Minenergo. I believe the 175 means 175 tons/hour, so that annual capacity is about 1.5 MT/year. The concept of the plant is described in *Elektricheskie stantsii* (1974:4, pp. 12–16). Stack gases from the power station are used as a source of heat, and the plant is to produce semicoke (some of which is to be processed into briquets), tar, and gas. The gas and tar will be burned in the power station. The goal of the exercise would thus appear to be to test the process on a larger scale than used earlier at Kalinin and to produce a supply of semicoke and briquets that could be used experimentally in different ways. One of the long-range ideas is to use the semicoke as part of the mixture in production of metallurgical coke. Though construction of this pilot plant seems to be the crucial next step in the commercialization of the energy-technological process, work is apparently not proceeding very fast. One source claims it is already in operation (Krapchin, 1976, p. 134), but Chukhanov says in 1976 only that it is being built (*Vestnik ANSSSR*, 1976:9, p. 120), and a story in *Sotsialisticheskaia Industriia* (May 1979) makes clear that it is *not* finished, and that in fact work on its construction almost completely stopped in 1979. Minenergo has responsibility for its construction, but has not given it a high enough priority to get it finished and into operation.

There seems to be another coal conversion process as well that could be used on Kansk-Achinsk coal, called thermal-contact processing. Actually there may be more than one variant of this process. One source speaks of thermo-coal produced in cyclone chambers using hot gas as the source of heat (Krapchin, in *Ekonomika i upravlenie ugol'noi promyshlennosti*, 1977:2, pp. 12–14), while another speaks of thermal-contact processing in an installation using a fluidized bed. The account here is based on a description of the cyclone version. The unit involved is described as the TKKU-300, which I believe means 300 tons/hour. This converter seems to have been developed under the auspices of Promenergoproekt, another of the Minenergo design organizations. Coal in small sizes (one source says 13 mm or less) is heated by contact with hot gas to 550°C and held at that temperature for 4–5 minutes—yielding as products from a ton of coal 550 kg of "thermo-coal" with a heating value of about 6,200 kilocalories per kilogram and 35 cubic meters of gas. The thermo-coal still contains volatile elements—about 40 percent. The cyclone process was developed by IGI in experimental stations at the Moscow coke-gas plant and the Khar'kov coke-chemical plant (Krapchin, 1977, p. 13). The fluidized bed process was developed by ENIN and the Eastern Coke-chemical Institute in Sverdlovsk and is embodied in a small pilot plant apparently located in Sverdlovsk (*Pravda*, 21 March 1977).

At one point it was being suggested that the next step in the coal conversion program should be the construction of a plant of 24 million tons capacity on the way to the construction of two 50-million-ton plants at a later date. The timetable is vague, but it seems even the 24-million-ton plant would not be built until after 1990 (*Elektricheskie stantsii*, 1974:4, p. 16; and *Energetik*, 1974:8, p. 37). The 24-million-ton project would use both the ETKh and the TKKU units. Other sources seem to favor going directly to a 50–60-MT/year plant using the TKKU equipment (*Pravda*, 21 March 1977; and Krapchin, 1977).

Finally, there is also a proposal for synthetic liquid fuel from coal, via complex processing with oil. It is apparently different from solvent-refined coal, and the main goal is to hydrogenate coal, using oil as the source of hydrogen (*Planovoe khoziaistvo*, 1977:8, p. 98). But this is still only at the theoretical and experimental stage; there does not appear to be any pilot plant proposal yet.

Some aspects of the research, development, and demonstration program for coal processing are difficult to understand. First, it seems poorly oriented toward solving the major problem of putting Kansk-Achinsk coal into a form that will permit it to be used like any other coal for energy purposes. In the original one-million-ton pilot plant, it is intended to use much of the output (the gas, the tar, and perhaps some of the semicoke) as power plant fuel at the Krasnoiarsk station, and this doesn't help with the crucial need to produce power plant fuel for the Ural and Volga regions. The economics of combining the process with local power generation (to use the gas) creates puzzles. The projected 24-million-ton plant would produce only 1.71 million tons of standard fuel in the form of gas, and this would cover only a relatively small power station—i.e., less than 1,000 MW. (A 1,000 MW station requires 2–2.5 million tons of standard fuel per year, for example.) The capital savings expected in the very large (4–8 GW) stations, in which direct burning of Kansk-Achinsk coal is planned, would probably offer strong economic arguments against this "energy-technological" line. One is left with the feeling that the effort is focussed on a technology not well suited to solving the main problem. And in fact some experimentation is being done with other methods of processing the coal that seem more relevant to the problem of making it possible to transport. One method in particular, the "autoclave method" (described in Krapchin, 1970), would use steam to dry and raise lump strength of Kansk-Achinsk coal to make it transportable. But tests on samples of coal from the Irsha-Borodinskoe field that an Austrian firm performed for the Russians concluded that this treatment would not suffice to solve the problems of decomposition and self-ignition. "Measures recommended by the firm for this purpose are expensive and have not been tested industrially" (*Ugol'*, 1974:9, p. 73).

V. A. Kirillin states that many methods of utilizing the coal will probably be used, and it is interesting that the Scientific-technical Council of Minenergo says that "until we can obtain an upgraded fuel from Kansk-Achinsk coal, we acknowledge that it will be possible to use untreated coal in the European part of the USSR, first of all in the Ural" (*Energetik*, 1974:8, p. 37). Some Kansk-Achinsk coal is, in fact, shipped considerable distances for energy use. The output from the mines at Nazarovo and Irsha-Borodinskoe exceeds local power station use, and, according to Ia. A. Mazover (*Planovoe khoziaistvo*, 1975:6, p. 79), they have been successful in shipping it 400–500 km. But the idea of shipping much untreated coal seems to me quite an overstatement of what is possible and an indication of pessimism about any early answer to the question of how to process the coal into more transportable forms. Mazover also says that it may be necessary to ship run-of-mine coal, using films and emulsions to protect it—i.e., to prevent its drying out and crumbling and undergoing spontaneous combustion (*Planovoe khoziaistvo*, 1975:1).

POWER TRANSMISSION

One method for connecting Eastern coal to the energy needs of the West is to build mine-mouth power stations at the major Eastern sources of energy coal (Ekibastuz and Kansk-Achinsk) and send the power west via ultra-high-voltage bulk transmission lines. Despite an established record of progress in this technology, the USSR will have to traverse a considerable R and D gap to create transmission lines suitable for this need; the distances involved require moving to higher voltages than those yet mastered. Because of the great distances the Soviet power industry must cope with, the USSR has had at various times longer transmission lines and higher voltages than other countries. At present the Siberian situation directs great attention to this technology and forces the USSR out to the frontier.

A rather detailed study of Soviet technological progress in high-voltage transmission concluded that

> the USSR moved by 1960 from a position in which it was a follower of technological trends in the high-voltage field to one in which it ranks among the leaders, both in AC and DC. This is not surprising to the extent that high-voltage technology is one area in which a country's performance as an innovator is to a large degree a function of its geographical and economic problems. . . . In the 1960s, however, the Soviet Union has in several respects lost its leading position in the development and diffusion of voltage technology. [Amann et al., 1977, p. 224]

The present goal in DC transmission is to step up from 800 KV lines to 1,500 KV. Beyond that, plans call for an eventual further increase to 2,200 KV. In AC transmission the goal is to move up from 750 KV to 1,150 KV.

The first step in getting Eastern mine-mouth power to market in the West is a plan to send power from the Ekibastuz region to the Ural region over an 800 KV DC line. There is already a line being operated at this voltage (Volga-Donbas), so this new line should present no great technological problem. To deliver power from mine-mouth plants in Ekibastuz to the power system of the Center Region at Tambov, a distance of about 2,400–2,500 km, however, it will be necessary to use a 1,500 KV DC line. The Ekibastuz-Center line in its present conception would have a capacity of 6 GW, and so could handle only part of the total output of the planned Ekibastuz generating complex of 16 GW. Much of the Ekibastuz power output will be used closer to home, especially in the Ural region.

From the Kansk-Achinsk power station complex it is planned to distribute power *within* Siberia using an 1,150 KV AC line, of which the first step will be a 500 km segment between Itat and Novokuznetsk. This will be a significant step beyond the 750 KV AC lines already mastered. But to get power to the Ural region from the planned mine-mouth plants in the Kansk-Achinsk fields, which at 4,000–4,500 km are much farther away than Ekibastuz, it will be necessary to employ something like a 2,200 KV DC line. This is a big increase even over the 1,500 KV voltage planned for the Ekibastuz-Center line still to be built. The Soviet power industry planners are understandably much more tentative about the prospects for the 2,200 KV line. The capacity of a 2,200 KV DC line is said to be 12–13 GW. It is further said that it would be necessary eventually to introduce a second line at 2,500 KV DC, which would have a capacity of 40 GW. Together, these two lines could handle a large fraction of the total power output from the Kansk-Achinsk complex of stations. The current development task is to create the 1,500 KV DC line and the 1,500 KV AC line—let us begin with a brief description of where development now stands on these two projects.

The Ekibastuz-Center Line

The plan for the 1,500 KV line from Ekibastuz to the Center has been on the agenda for quite a few years. It was said at one time that the line was to be constructed in the 9th Five Year Plan period (1971–1975)—*Izvestiia VNIIPT*, No. 16, 1970, p. 17). But when the 9th Five Year plan was promulgated it did not actually specify a target regarding construction of the 1,500 KV line or even production of equipment for it. Apparently only experimental and design work for such a line was carried out in 1971–1975. The concept for the project is laid out in *Energetik* (1971:3, pp. 8–10), and the article in Izvestiia VNIIPT cited above said that an experimental line at 1,500 KV was put into operation "several years ago" (*Izvestiia*, 2 February 1974). I suspect that this was only an experimental section at VNIIPT—the article in *Izvestiia* cited above shows a picture of a test section on which they are testing for corona and radio interference. Kirillin reported in 1974 to the annual meeting of the Academy of Sciences that work on the Ekibastuz-Center line was underway. Popkov says in an article in *Pravda* (9 October 1974, p. 3) that the design for the line, completed over a year earlier, was still sitting in the files of the expert commission. Popkov also said that 60 items of new *equipment* had been designed for it. When we get to the guidelines for the 10th Five Year

Plan (1976–1980), no mention is made of constructing the 1,500 KV line, but the guidelines say that it is necessary to master the production of equipment for it (*Osnovnye napravleniia*, 1975, p. 36). The target date for its completion has therefore shifted from sometime in the first half of the seventies, to sometime beyond 1980—i.e., by at least 10 years. There may have been problems in producing the equipment, or perhaps the R and D work done so far has failed to convince people that the proposed design will work. A report in 1977 says that testing of the equipment is proceeding (*Sotsialisticheskaia Industriia*, 2 October 1977), and it is claimed in a story in *Soviet News* (12 June 1979) that work has started on the line.

There is still some time for getting 1,500 KV technology perfected. The 10th Five Year Plan envisages the completion and commissioning of the first 500 MW unit in the first power station of the Ekibastuz complex before the end of the period. A story in Pravda (16 March 1975) mentions the end of 1978 as the target date. A start on the construction of the second of the projected four power stations in the complex is also intended in the 10th Five Year Plan. In the Kansk-Achinsk complex, the language of the guidelines does not suggest completion of even the first station (the Berezovo station) during 1976–1980, but only progress on its construction. Thus the only power available for interregional transmission by 1980 will be from the Ekibastuz station, if it is in fact commissioned and begins to operate successfully.

The Siberian Line at 1,150 KV

Progress on the 1,150 KV AC line appears somewhat better. The 9th Five Year Plan said that a section of this line connecting the Krasnoiarsk and Kemerovo systems should be constructed by 1975, so that this section could be used for settling design decisions and for accumulating experience (Baibakov, 1972, p. 102). The guidelines for the 10th Five Year Plan (1976–1980) say that the production of equipment for this line is to be "mastered" (*Osnovnye napravleniia*, 1975, p. 36) and that the line is to be constructed (ibid., p. 25).

As yet, however, there is no reasonably confident and convincing answer as to how power from Kansk-Achinsk stations will be transported to the European USSR. I have seen no indications of significant development work on 2,200 or 2,400 KV DC equipment or line design. There is no mention of any of the series of preliminary design studies that would be performed for such a project (*avanproekt* or *eskiznyi proekt* for example) nor any statement of intermediate R and D targets such as experimental work with prototype equipment. Some authors even question the technical and economic feasibility of moving up to that voltage level. There is a great deal of discussion and experimentation on longer-range answers to the problem of long-distance bulk transmission, such as cryogenic lines, waveguides, and gas-filled conductors. Research work is being done on these. None of the commentators seems to think seriously that these approaches will be developed soon enough to get the power intended to be produced at the Kansk-Achinsk complex to the European USSR, but they may not be really any more unfeasible or distant than 2,400 KV DC technology. One important feature of the Soviet situation is that there are very large flows of power to be transmitted, and there may be economies of scale with superconductive or cryogenic approaches that would give them a competitive edge over traditional overhead lines when the size of the flow is so large.

To conclude, the Soviet literature on progress toward the higher voltages in power transmission is not very informative. A priori, these would seem to be interesting cases for illuminating how the Soviet R and D system operates on the frontier of some technology. Unfortunately, the strongest impression one gets from the rather slim accounts of what is being done, is one of hesitation and vacillation, without much idea of how they evaluate alternatives and make decisions about how to move forward. This uncertainty may derive in large part from the uncertain state of the Kansk-Achinsk plans themselves. As indicated in Chapter 3 and in the sections in this chapter on coal processing and slurry pipelines, it is still not clear that it will be feasible to burn this coal in mine-mouth power stations or that there may not be some equally attractive alternative way to move the energy in it to the West. In that situation it is not surprising that there should be some hesitation about making large commitments to move on to actual development of the technology for transmitting this energy by wire.

MAGNETOHYDRODYNAMIC GENERATION

The Soviet interest in magnetohydrodynamic generation of power (MHD) began rather early. According to a careful and detailed Western review of Soviet MHD work, "successful feasibility demonstrations in the United States in 1959 provided V. A. Kirillin . . . and A. E.

Sheindlin with the rationale to initiate an MHD-research development program in 1961" (Rudins, 1974, p. 3). Research was first pursued intensively at the Institute for High Temperatures at the Moscow Power Institute (Moskovskii Energeticheskii Institut or MEI). From the very beginning MHD apparently had the personal support of V. A. Kirillin, who was at one time the Director of the high temperature laboratory from which the Institute was formed, although the major figure directly in charge of the program has been A. E. Sheindlin, Director of the Institute. The Institute was transferred to the Academy of Sciences in April 1967, and the work expanded to a more ambitious scale. Other institutes have been involved as well. There is a reactor at the Krzhizhanovskii Power Institute (ENIN), and the Kurchatov Institute has also been involved. In 1964–1965 the first pilot unit designed to demonstrate the actual working of the process was completed in Moscow; it is said to have been in continuous use since. It had a theoretical capacity of 200 KW. The results obtained with this experimental installation were sufficiently promising that a decision was made to create a full scale integrated pilot facility, the U-25, started in 1966 and completed in the spring of 1971 (*Electrical World*, 15 August 1973, p. 25). There is also a small pilot plant in the Ukraine (the Kiev-1) but little discussion of its purpose or status can be found. An article in 1974 speaks of the need to master and put into operation the MHD installation at the Kiev GES-2 station, and this may be an indication that the Kiev MHD unit was technically a mistake or was based on an idea later discarded. At any rate, there is a hint here that not everything in this program has gone smoothly (*Energetika i elektrifikatsiia*, 1975:1, p. 54).

The Soviet interest in MHD has usually been understood in terms of its being the most promising method for significantly increasing the efficiency of electric power generation. As explained in Chapter 1, electric power is a disproportionately large consumer of fuel in the USSR, which has led to a great concern with technologies for reducing the heat rate, such as cogeneration and the combined cycles discussed in Chapter 3. But the arguments for MHD seem to have varied over time. The development program started and is being carried forward using natural gas as a fuel, with residual fuel oil considered as an alternative for the first industrial plant, though these are now precisely the fuels that must be replaced with others in the European areas where use of MHD generators is intended. For some time the Soviet sponsors of MHD research have known that this will become a viable technology only when it is converted to using coal, but a tremendous inertia carries the program forward in its gas-based form. Also in the beginning, it was thought that MHD would be a technology employed primarily in base-load generation. The reasoning behind this expectation was that, as a fuel saving technology with high unit capital costs, it should be used a large number of hours per year—5,000–6,000 (Kirillin, 1974a; available as JPRS 62–139). But the thinking has now shifted to a conclusion that MHD should be used in TETsy and as a cycling technology. Used in TETsy, it would permit the burning of sulfurous oil (sulfur will be removed by an interaction with the seed and recovered in the regeneration of the seed material) in a way compatible with urban pollution standards and would also help to satisfy the need for generating technologies with rapid start-up capability. These are important points in the justification of the design for the first industrial plant presented at the joint U.S.-USSR 1976 colloquium on MHD in Moscow (Kirillin, 1976, pp. 27–32). As explained in Chapter 3, the need for equipment that can be started up on very short notice is acute (especially in view of the increasing emphasis on nuclear power plants, which are essentially base-load plants), and so this is a good selling point.

The current status of the program is that enough has been learned from the operation of the U-25 testbed facility to make the construction of an industrial plant appropriate, and such a plant is now being designed. Several years ago it was predicted that this first industrial unit would be an 800–1,000 MW block, presumably in line with the idea that it would be a base-load plant (Kirillin, in *Vestnik ANSSSR*, 1975:2, p. 12). But the most recent statement indicates that the first plant will be built as a heat and power combine, probably in connection with an already existing heat and power combine near Moscow, and will use natural gas or oil (Kirillin, 1976, p. 29). There are several statements that reflect hopes that such a unit can be built before 1985: a representative of Teploenergoproekt said at one point that this plant must be ready sometime in the 1981–1985 period, and as recently as 1977, the Minenergo journal says that it is possible for it to be commissioned midway through the 11th Five Year Plan (1981–1985—*Elektricheskie stantsii*, 1974:4, p. 8; and 1977:7, pp. 13–14). The change of plan to employ the first MHD unit in a heat and power installation probably reflects in part the availability of a suitable turbine (i.e., the 250 MW supercritical extraction turbine), and the size of the project

is determined starting from the technical characteristics of that turbine.

One of the most interesting features of the MHD program is that
it is being carried out with a considerable amount of cooperation with
the United States. According to Rudins, MHD is an area with a long
history of international cooperation, and the Soviet Union has been
associated with the formal and informal international groups working
in this area for a long time (Rudins, 1974, p. 1). The United States
was an early pioneer in MHD research, but development languished
in the sixties, when many were skeptical that this could be a competitive technology. Rising energy prices, a desire to increase coal as a
power station fuel, and the need to deal with the pollution problem
posed by coal later gave MHD a renewed high priority in the United
States. At this point, the Soviet and U.S. efforts were quite complementary, and it was easy to argue that a joint effort would bring big benefits. The United States had done a great deal of design work on testing
of concepts, components, and materials and had the superconducting
magnet technology. The USSR had followed a different R and D philosophy (about which more below) in setting up an expensive pilot
plant intended to provide a testbed for testing individual elements
and the system as a whole in a functioning plant. MHD was specified
as a field for cooperative research in the U.S.-Soviet agreement on
scientific and technical cooperation signed in 1972, and this effort has
since expanded to extensive and real cooperation. The United States
has supplied a superconducting magnet for use in the U-25 facility,
which by the end of 1977 had been installed and cooled and was ready
to start operation (*Department of Energy Information*, 21 October
1977). We have also designed several channels for testing in the facility. The Russian contribution is to provide access to the U-25 testbed
as a facility for testing different design ideas. The USSR side actually
conducts the jointly planned tests and is to share with the U.S. side all
the experimental results. One of the by-products of this cooperation is
that the U.S. side has experienced a much more intimate involvement
and insight in a Soviet R and D program than is customarily possible,
and the American participants have provided an interesting commen-
tary on some distinctive features of Soviet R and D.

The first of these is a striking difference in R and D philosophy.
The American approach to MHD has been to do extensive work on
components and concepts and to accumulate enough data so that the

planners can feel they understand the process well enough to assure that a commitment made to an experimental facility will work as designed. This is a philosophy, moreover, which seems to be true of U.S. R and D in general. The Russians have gone much more directly into the creation of an expensive plant, in which various elements of the whole system can be experimented with, redesigned, and upgraded within the context of the system as a whole. This approach has its costs. The Americans are much impressed with how expensive these facilities have been. One of the points in selling the cooperative program to U.S. authorities has been that the U-25 facility would cost about 100 million dollars for us to build and that the outlay for such a facility is not something the U.S. program could justify on its own. They also comment on the disadvantages that come from fixing designs in concrete and hardware at too early a stage in the game. In reporting on what has been gained from the cooperation, the U.S. side has said that "we have learned of Soviet difficulties experienced in trying to develop MHD with a rigidly designed pilot plant" (U.S. Congress. House, 1975, p. 550). Rudins has an interesting suggestion (1974, p. 24) that one motivation for the Soviet decision to build the U-25 plant was the need to ensure visibility for the program and a desire to get enough momentum and investment to make it difficult for the authorities to terminate the program. He also suggests that one consequence of this approach is that, in its components, the design of the plant could not be technologically venturesome and had to stick very close to established technology (ibid., p. 39).

Another conclusion from the MHD experience is that the Soviet R and D establishment can be quite bold in a commitment to far-out ideas. Apparently support for the MHD effort has continued without a setback over what is now almost two decades. It has been very expensive. Rudins attempts an estimate of the total costs of the MHD program up to 1974 as 250 million dollars (ibid., p. 73), though he does not explain how he obtained this figure. There have also been great delays. Nevertheless, the support has continued undiscouraged even in the years when the United States was weakening its support for MHD. Sheindlin is quoted as saying that "it was not easy for us to stay in the field in isolation, but we were sure of the correctness of our approach and continued it" (as reported by Robert Toth in the *Louisville Courier-Journal*, 6 April 1975, p. E9). This is a different conclusion from what we have found in some other cases, where what is being done in the rest of the world technologically operates as a powerful argument in determining the directions of Soviet R and D. There are probably several factors explaining this commitment. One is that Kirillin became head of the State Committee for Science and Technology (GKNT) and a Deputy Prime Minister, and so was in a uniquely favorable position to turn his personal belief in the project into an institutional commitment. Rudins says that Kirillin deliberately chose MHD as an example of how Soviet scientific development could be taken to the applications stage (1974, p. 73). Furthermore, MHD seems to have consistently had the support of Minenergo. The degree of involvement is indicated by Sheindlin in his formulation that the test facilities were "created by Minenergo, under the scientific supervision of the High Temperature Institute" (*Vestnik ANSSSR*, 1968: 11, p. 20). Elsewhere he says the U-02 and the U-25 were "developed jointly by Minenergo and the Academy of Sciences" (*Izvestiia*, 27 September 1976, p. 5). Rudins reports that Soviet officials involved with the program say that one of the distinctive features of the system, its embodiment in a plant that actually delivers power to the Moscow network, is a design decision made in part to keep Minenergo convinced that this is a commercial technology (1974, p. 34).

The MHD history reinforces our general conclusion about Soviet R and D: that initial plans are always overambitious. Kirillin was saying in 1974, at the annual meeting of the Academy of Sciences, that the first demonstration plant would have a capacity of 1,000 MW and could be developed by 1981 (*Soviet News*, 10 December 1974). Some of the earlier accounts indicated that the next step to follow the U-25 would be a 500 MW demonstration plant (*Elektrichestvo*, 1970:4, p. 5), and it is interesting that the actual proposal has now moved back to that size plant. Perhaps this reflects the influence of the interaction with the Americans, though it seems more likely that a plant on this scale will be simply more feasible to construct. The proposal for a 1,000 MW plant was based on a presumption that there would be a 500 MW turbogenerator unit available by the mid-seventies (Kirillin, 1974a, p. 11), and the delay in getting such units mastered may have changed their minds.

The whole history also illustrates the kind of twists and turns an R and D project undergoes as it advances. As explained, the conception of the role this technology should play has been substantially redefined—it is not clear how best to interpret this. I am prepared to believe that the original decision to base the early development stage on gas as fuel was sensible enough at the time and that the preferability of using MHD as a coal technology only emerged when the fuel balance environment changed. Indeed, it may still turn out that gas is the best fuel for the applications they now intend for MHD—i.e., in the European SSSR, where gas may well be no more scarce or expensive than either coal or oil. In fact, what should be questioned perhaps is the current assessment that it must ultimately be adapted to coal. Given the transport considerations, the biggest expansion of coal-fired plants will be based on cheap, low-quality coal, mostly in the East, and one wonders if the fuel savings in such a situation will be worth the capital cost. The real dilemma for this technology is that, for the regions with high-cost fuels, nuclear power seems to have the competitive edge for base-load uses. In these regions, oil and gas will be reserved as fuel for peaking and intermediate uses; the competitive position of MHD in this area is hurt by the low number of hours during which its big advantage—lower fuel expenditure rates—is being exploited (all this is detailed in Kirillin, 1974a). The Soviets argue that it is possible to develop MHD to a commercial stage by continuing the gas direction and then modify the technology for coal. But, says Rudins, "some noted U.S. MHD specialists believe just the opposite: that coal plants are totally different from natural gas plants" (1974, p. 75). The shift from justification in terms of fuel economy to justification in terms of environmental advantages and suitability for intermediate load generation is reasonable enough, though the rearrangements seem just opportunistic enough to make one wonder how firmly grounded they are in cost benefit calculations.

Finally, the USSR seems to carry its preference for early commitment, loosened by flexibility and insurance in design, right on through succeeding levels of demonstration facilities. The Soviet paper on the considerations guiding the design of the first industrial MHD plant, given at the joint symposium in Moscow in 1976, shows a great concern with providing expensive supplementary facilities that will permit the facility to work under unexpected adverse circumstances that may flow from the early commitment to a design that has not been oriented to a well-specified environment on the basis of refined knowledge. This paper states that

> in order to accelerate the adoption of the new equipment, the plan provides certain measures designed to facilitate the attainment of rated indices. These will make the facility somewhat more expensive and subsequently it will be possible to discard them. The essence of these measures consists in their ability to compensate for possible miscalculations

that may have occurred in the designing of individual pilot units of the facility and to preserve the capability of the unit to operate while the mistakes in designing are being corrected. [Kirillin, 1976, p. 31]

The paper proceeds to list a considerable number of such provisions. A desire for flexibility to cope with technological surprise is a perfectly rational decision, of course, but against the background of the many other examples in which plants are built ahead of the technology, it is easy to imagine that the managers of this program are taking considerable risks.

GEOTHERMAL ENERGY

There is a long-standing Soviet interest in geothermal power. Exploration drilling for a power station at Pauzhetsk on the Kamchatka Peninsula in the Far East began as early as 1957 (Dvorov, 1976, p. 71). (Other sources say 1958—Kruger and Otte, 1973, p. 42.) The objective of that exploration was apparently to decide whether a sufficient flow of fluid could be developed to supply a power station. It was thought that a start could be made with a small station, which could later be expanded if more resources were discovered in the process of exploration. The work was conducted by the ANSSSR and by Mingeo, so I suspect that the project was motivated more by basic research concerns than by a desire on the part of Minenergo to find a way to utilize new primary sources for electric power.

In September 1961, the Presidium of the ANSSSR established a Commission on Hydrogeology and Geothermics (Dvorov, 1976, p. 24), which was assigned the mission of coordinating and leading basic research in this area.

The most significant milestone in attention to geothermal sources was a 1963 decree of the Council of Ministers on the development of geothermal resources. The tone of the decree is impatient—as if plans had been around for some time but that action had been delayed. It is an action document, establishing timetables and allocating responsibilities, and contains both a development program and an outline of further basic research directions. The Academy of Sciences of the USSR was given responsibility for organizing basic research, and in January 1964, it transformed the Commission mentioned earlier into a Scientific Council on Geothermal Research. (Dvorov is *uchenyi sekretar'* of this Council.)

Things then began to happen, but slowly. On the basic research side, it seems that the most intensive activity was the exploration of resources. The oil and gas exploratory organs were instructed to include geothermal data in their studies, and there are numerous reflections of their studies and concern with geothermal resources in the literature of the following years.

On the side of experimental and development work for utilizing these sources, work went ahead on building the experimental power station at Pauzhetsk using direct steam. Construction is said to have begun in 1964, according to a design developed by Teploenergoproekt (Kruger and Otte, 1973, p. 42), and the station was "commissioned" in 1967 (Dvorov, 1976, p. 26). It was originally intended that this station would be a 12 MW unit (Zolotar'ev and Shteingauz, 1960, p. 177), but in the end it was given a capacity of only 5 MW in two separate 2.5 MW units. These were probably standard turbines—Dvorov says they were MK-2.5 models produced by the Kaluga plant (1976, p. 73).

The Pauzhetsk station was to be followed by a second one at Makhachkala on the western shore of the Caspian Sea—also with a planned capacity of 12 MW (*Elektricheskie stantsii*, 1962:1, pp. 3–4; and the 1963 decree). But this part of the program was apparently never carried out.

The Pauzhetsk facility was a fairly simple one. The resource was superheated water, and on extraction, steam was separated from the fluid and was sent directly to the turbine. Some impurities were carried over with the steam, but it is asserted that they did not damage the turbine.

The Pauzhetsk station was reasonably successful and operated more or less continuously after it was built (Dvorov, 1976, p. 73). But Dvorov's book contains a table indicating that, although its nominal capacity was 5.4 MW, its "disposable" capacity was only 3.2 MW. Another source explains that the parameters of the steam at the site turned out to be lower than estimated, so that the station could not develop the planned capacity (Zhimerin, 1978, p. 186). On the other hand, it is claimed that the capacity of the station is now being expanded to 10–15 KW (Dvorov, 1976, p. 74), which may be a matter of developing a larger supply of steam. Dvorov also gives investment figures that work out to 906 rubles per KW, which is 8–9 times the cost per KW usually cited for most power stations.

More or less contemporaneously with the Pauzhetsk station, an experimental secondary-fluid station based on Freon-12 was created to use the lower-temperature waters of the Paratunka geothermal field, also on the Kamchatka Peninsula. This plant is said to have begun operation in 1968, about a year after the Pauzhetsk station. I suspect that Minenergo had even less interest in that station—the design work for it was done by an Institute of the Siberian Branch of the Academy of Sciences and the Institute of Refrigeration Machinery, and the construction of the station was carried out by an organization in Minneftekhimmash.

The Paratunka station apparently worked so badly that it was soon abandoned. It is usually not even mentioned today; Dvorov rationalizes that it was really only a station-laboratory rather than a power station and says that although it functioned for a while, it has now been dismantled so that the freon equipment can be redesigned to raise its efficiency. P. S. Neporozhnyi described in 1972 what seems to be this same equipment but spoke of it as being installed in the Shatura power station near Moscow, and I imagine that it was moved there for further experiments after being dismantled at Paratunka.

It appears that further development of geothermal power stations will now take a different direction from the original steam concept. Several additional projects more or less like the Pauzhetsk plant were proposed (for Petropavlovsk-Kamchatskii and the Bannaya Valley on Kamchatka and for Kunashir Island in the Kurile Chain), and the 1963 decree directed that series production of equipment for such stations be started. But these were never built, and nothing is heard of such projects now. Most discussions now suggest that the supply of heat in the form of natural steam is so small that it will not be feasible to develop more stations of the Pauzhetsk type.* And the absence of any follow-up discussion on the secondary-fluid approach, which could utilize water at lower temperatures, makes me think that experiments involving the freon equipment were abandoned or were too disappointing to justify use of that approach. Specialists engaged in basic research in the field have always said that if extensive use is to be made of geothermal power the crucial step is to solve the problem of massive heat transfer from hot rock at great depths to generate steam. Some of the research that has been done follows these lines, one of the proposals being to create a cavity at great depth in hot rock with a nuclear explosion and circulate water through it. Recent discussions

*One current source *does* assert that the Petropavlovsk-Kamchatskii station is being built and that construction has begun on a 300 MW station (Alekseev, 1978, p. 169). I doubt that the author knows what he is talking about.

mention a project for a larger power station "of the circulation type" to be located in the Makhachkala area, and an article in 1978 says that construction of such a station in the USSR is "contemplated" (*Elektricheskie stantsii*, 1978:5, p. 8). Nonpower uses of geothermal heat are a way in which substantial fuel savings could be obtained by using water at temperatures well below boiling. This kind of use is growing moderately in the USSR but faces many obstacles, including institutionalized skepticism.

Because development of geothermal sources has not worked out well, the history of development is not very fully described in the literature, but all the evidence available suggests that little has been accomplished to develop the technology required to bring this source into the fuel balance. The Ministry of the Gas Industry gives some output figures which suggest that the output of water and steam has *fallen* since the early years of the program (*Gazovaia promyshlennost'*, 1977:6, pp. 19–21).

The explanation for slow progress in doing the research and developing the technology to use geothermal power is that the task has never had anything more than halfhearted leadership. As one exasperated advocate says, "there is today no clear line defining a program for geothermal resources" (*Ekonomicheskaia gazeta*, 1973:48), and most Soviet sources would add that there is no organization with adequate interest and responsibility for pushing it. The division of responsibilities is: basic research to ANSSSR; development of resources to Mingaz; utilization to customer ministries such as those for agriculture, for the chemical industry, and Minenergo. The Ministry of the Gas Industry has charge of drilling the wells and producing the water and steam, which it does today through four field administrations. But it has never really had at the top level any serious interest or commitment to geothermal resources, which are after all a minor sideline to its main mission of finding, producing, and transporting natural gas. At one point the Ministry simply liquidated its department in charge of geothermal work, according to a critical analysis published in *Pravda* (15 January 1974). The article says that drilling for geothermal resources was not protected by having a separate allocation in the Ministry's budget or a separate target in the plan. One can easily imagine that the Ministry preferred to use all its drilling capacity, probably the greatest bottleneck it faces, for its main task of finding and producing gas.

Nor have any of the potential users of geothermal resources done much to promote geothermal development. Minenergo officials are consistently on record as saying that geothermal energy is irrelevant to the problems of large-scale power generation. One of Minenergo's project-making institutes designed the Pauzhetsk plant, but the equipment was developed outside the ministry, the plant was built by another ministry, and I have not been able to find any evidence that it is even under the jurisdiction of Minenergo, whose relationship to the Paratunka plant is about the same.

A review in the Minenergo journal *Elektricheskie stantsii* (1978:5) reports on what is apparently a serious effort within the Ministry to evaluate the potential of geothermal resources for power generation. The author explains that use of water at 150°C and below is uneconomic anywhere in the USSR, and even when water at 250°C and above is available it cannot compete against other sources unless it is at relatively shallow depths to keep drilling costs within reason. Overall, the author confirms the generally pessimistic views about geothermal power said to be held in Minenergo. As for other potential users, it is reported that the Ministry of the Chemical Industry has no interest in the brines as raw material sources, and that the Ministry of Agriculture drags its feet on using geothermal resources for heating greenhouses and other agricultural applications.

Beginning in the mid-seventies there has been renewed pressure from the center to give R and D for renewable sources a higher priority. (See, for instance, a plea for this change in *Kommunist*, 1976:2, pp. 62–65.) The gas industry has been repeatedly criticized for its failure to carry out its responsibilities for geothermal energy and now seems to be treating this area with a higher priority. An article in the Mingaz journal laid out an extensive program of expanded R and D work on geothermal problems. Mingaz is drilling a few more wells than in the past, planning for some growth in output of geothermal waters, and doing much more research on such problems as corrosion and the problems of drilling in these difficult environments (*Gazovaia promyshlennost'*, 1977:6, pp. 19–21). This may be more an effort to simulate some activity than a genuine commitment to a higher priority, however, and the complaints go on. A recent article expresses great skepticism:

> The country has an organization responsible for the utilization of deep-earth heat. Fifteen years ago this responsibility was assigned to Mingaz. It created field administrators for this purpose, relatively small and poorly equipped agencies one must admit. This essentially ended the min-

istry's concern with the accomplishment of a task of great national economic importance. Not much exploration and prospecting is being done for hot water and steam. The ministry shows no concern for the designing and manufacture of specialized equipment capable of operating dependably at high pressures and temperatures. In short the ministry, engrossed in its principal business, treats geothermal engineering like a stepchild for whom it lacks sufficient love, time, and money. [*Pravda*, 15 December 1978]

TIDAL POWER

The Soviet Union has large tidal power resources, most notably on the Arctic coast, but also in the Sea of Okhotsk in the Far East. The capacity of some of these sites is said to be very high. The figures one finds quoted are quite variable, no doubt depending on how large an area is assumed to be feasible or economic to block off. One source says that the capacity in the Mezen location on the Arctic coast, for example, could be 14 million KW (*Elektricheskie stantsii*, 1962:1, pp. 3–4). That is equivalent to three Bratsk projects or seven nuclear power stations like the one that has been built near Leningrad.

There appear to be two major issues in tidal power that govern the direction of R and D efforts. The first is the problem of building the elaborate structures required, in the inhospitable conditions of their locations in the USSR, at a cost that is acceptable. The second is the exploitation of its potential within a power system, since tidal power has a peculiar time profile, with some possibilities of modification via pumped storage. Extensive studies are required that evaluate the economic feasibility of tidal power projects in connection with the optimization of their design and the nature of the system into which they would fit.

The Soviet development program has thus far been oriented mostly to the engineering problems, via the building of a small prototype station at Kislaia Bay on the Kola peninsula, near Murmansk. The approach adopted to cheapen the cost of construction was to prefabricate the major structure elsewhere (in a dry basin later opened to a river), float it to the site, sink it onto a specially prepared bed, and then complete the barrage across the mouth of the inlet with an earth-filled dam. Work on the project began somewhere around 1964 (*Energeticheskaia, atomnaia, transportnaia i aviatsionnaia tekhnika. Kosmonavtika*, 1969), and construction took several years. According to one account the power house was in place by 1968 (Shabad, p. 34), another suggests 1969 (L. Bernstein, in *Civil Engineering*, April 1974, p. 47). Bernstein, chief engineer of the project, adds that this station first generated power in 1970. The unit uses a very small (400 KW) bulb-type, reversible, generating-pumping unit, supplied by Neyrpic, the French company that produced the units used in the French tidal power station.

Here, we are interested primarily in the R and D process, and although not a lot is said about the evolution of this project, it has been possible to gather a few interesting facts. The project was developed by Gidroproekt (which is under Minenergo), so that I believe this experiment has had its major support and direction from that source. Perhaps it is not irrelevant that the Minister of the electric power industry, P. S. Neporozhnyi, is a construction man. It is also interesting to find that the concept has changed considerably in the process of development. An early account said that the station would have a capacity of 1200 KW—i.e., presumably using three generating units (*Energeticheskaia, atomnaia, transportnaia i aviatsionnaia tekhnika. Kosmonavtika*). A later description described it as having *two* 400 KW units (*Energetik*, 1967:3), and as indicated, the completed project had only *one* 400 KW unit. Comparison of the drawings of early mock-ups with what the structure looks like in place suggests that the whole structure was reduced in size, though Bernstein says that the prototype structure is that of a full-size station, and it might be that only some of the generating equipment was eliminated.

The explanation for this scaling down of the project could be some technical problem—they may have had to change their ideas of how big a structure it would be possible to manage in the towing and seating phase. It seems equally likely, however, that there was strong pressure to cut the costs of the project as it evolved, both in terms of total investment and in terms of foreign exchange cost. In other words, even if the Soviet R and D system shows an undeniable inclination to support speculative and long-range directions in technology, we might well conclude that such projects are not immune from budget pressures.

Virtually nothing more has been published about operating experience with this experimental station, except that it is said to have demonstrated the success of the idea and that design work is proceeding for the construction of a much larger station at the Mezen location which, it is asserted, will have a capacity of 6 GW (*Soviet News*, 10 November 1970; and Bernstein).

This once again shows the usual Soviet preference for very large upward steps in the development of a technology. If such a station is actually attempted that would be a scale-up factor of 15,000! A project of 6 GW capacity would have to use either larger generating units (the largest reversible units the USSR has attempted to produce so far are the 200 MW units for the Zagorsk pumped storage station—the French station uses 10 MW units) or an incredible number of small ones. There are other, smaller sites available they could work on. But "design work" in Soviet parlance can cover a wide spectrum of studies, and the statement cited should not be taken to mean that the construction of such a station has actually been decided on. The other aspect of the development problem—evaluation and design to see how such a station would be justified in the growth of the system—must be the next major step. It appears that Soviet interest in tidal power is based as much on its potential help in the peaking problem as on its possible contribution toward expanding primary energy sources. Tidal power, because of its peculiar time profile, is not in competition with baseload technologies such as nuclear power. The future of the project thus crucially depends on what happens to competing technologies for peaking purposes, such as the various thermal forms of peaking equipment and pumped storage. That must be one of the major issues at the present stage of R and D on tidal power.

SOLAR ENERGY

Solar energy has long had supporters in the USSR, and some serious work on this technology goes back several decades. It has never been a "big-science" effort; nor has the central machinery pushed it seriously. Solar energy R and D has been performed in the republican academies, in small units within larger institutes both of the Academy system and the ministerial system, and in institutions of higher education.

The Armenian Academy of Sciences and the Power Institute (ENIN) cooperated for a while on a proposal for a power plant and built a solar house, but that program broke up in the sixties and experiments ceased. ENIN had a helio-laboratory in Tashkent, and ANArmSSR had a commission on heliotechnology, which later was absorbed in the Power and Hydraulics Institute (Institut energetiki i gidravliki) of the ANArmSSR. There is still some research on solar energy at ENIN.

The scientific journal for this area of research—*Geliotekhnika*—is published by the Uzbek Academy of Sciences, which includes the S. V. Starodubtsev Physico-technical Institute in Tashkent, which seems to be a major locus of solar energy research. It has a *geliotekhnicheskii poligon*, though no description of its facilities is available. A survey of the institutions in which the articles in the journal originate suggests that many have solar research as a sideline to their main activity, such as institutes of electronics (converters) and institutes doing design work where remote power sources are important (Tashkent communications institute). The Physico-technical Institute (Fiziko-tekhnicheskii Institut) ANTurkSSR works on solar cells. An article reviewing 10 years of the journal's activity says that "there has been a considerable expansion in solar-technology research in the Soviet Union in the last ten years," and adds that in addition to institutions active in 1965 (Leningrad, Moscow, Erevan, Tashkent, Ashkhabad), new schools have been established in the Ukraine (concerned with the effect of high-intensity light and heat fluxes and high temperatures on matter), in Moldavia (development of semiconductor photocells and effect of light on seeds), and in Azerbaidzhan (development of autonomous sources of supply using solar energy). They mention a three-meter solar furnace at the Uzbek Academy and a 10-meter furnace being built at Erevan. Their summary is that in 1975, there were 30 doctors and 100 candidates doing research on solar energy. It is thus clear that this is a very small effort and is more concerned with basic research than with development.

If this line of research has had little support (as indicated in our survey of R and D spending) or little direction from above, the reason is easy to see. The makers of energy policy make no bones about their conviction that solar energy is of no interest for current energy strategy, because it cannot compete with other alternatives. All the major figures in the energy R and D "establishment"—Academicians Kirillin, Styrikovich, Aleksandrov, and lesser lights in the ministries—say that the exotic sources are not promising for making a contribution to serious energy problems. For the most part, they ignore such sources in discussions of the energy future, but when they do feel called on to evaluate them, the following statement by Kirillin about solar power is typical:

> At the present time one can hear not a few statements that it is necessary to expand work aimed at utilizing the huge resources of solar energy and underground heat for the needs of large-scale power (*bol'shaia energetika*). However, thus far, no suggestions acceptable from an economical and technical point of view have been put forward as to how to do so. The difficulties in the use of solar energy on a large scale for

obtaining electric power lie in the fact that the technical-economic indicators of both presently realizable methods—concentrating the sun's rays
for heating a steam boiler, or using solar cells—are very low. [*Kommunist*, 1975:1, p. 52]

A recent article in the Minenergo journal dismisses hope for use of
solar cells as completely unreal. The author says that the cost of such
cells at present is about 100,000 rubles per kilowatt, and although he
is willing to concede that costs can be reduced by one order of magnitude, the prospect is that even after the considerable effort such a
program would require the cost to remain two orders of magnitude
above the 100 rubles per KW required to make solar cells competitive
with conventional sources (*Elektricheskie stantsii*, 1977:7, p. 9).

Kirillin and other authorities who influence energy R and D directions are thus prompted to conclude that any effort on these exotic
sources now should be directed toward long-range studies attacking
fundamental issues, and that, until such efforts succeed in making
some kind of breakthrough, solar and other such sources are not candidates for applied development effort.

The main exception granted is that a few special situations may
make solar energy competitive with conventional sources even today.
The major such uses involve small-scale autonomous power supply in
desert regions for pumping and desalination of water; irrigating pas-
tures and watering stock; supplying power for communications and
cathodic protection of pipelines; water heating; and supplementary
space heating and air conditioning for housing. A seminar held in
Ashkhabad in 1975 called for this to be the main direction in the near
future (*Geliotekhnika*, 1976:8, pp. 72–73).

It has been suggested that, in the United States, solar energy is likely
to make its way into practical use, less as a result of the big ERDA-
type programs such as the solar-tower power plant, than through the
efforts of private initiative concentrating on small-scale differentiated
local uses ("Solar Energy Research: Making Solar after the Nuclear
Model," *Science*, 15 July 1977, pp. 241–244). The direction being
taken in the USSR seems to embody that approach, though it seems
likely that in the USSR the forces substituting for attention from the
center are rather weak. Several of these small-scale technologies are
said to have passed the development stage, and some decisions have
been made to move into the production phase. One of the "programs"
in the R and D plan for the 10th Five Year Plan is for "bringing re-

newable resources into the fuel and energy balance" (*Kommunist*, 1976:2, p. 65). The Ministry of the Water Economy has set up an NPO* to design the equipment and establish production methods, and a plant is being constructed in Uzbekistan for routine production of equipment for solar power (*Kommunist*, 1976:2, pp. 63–64). But significant progress will require parallel initiatives in many other ministries as well, and the authors of the article just cited express the view that little will happen unless pressure is applied from the top to force planners in the Central Asian regions (where most applications are envisaged) to establish the enterprise to produce the equipment.

It is also possible that there will be some higher priority for solar R and D in reaction to the large spending under the U. S. program, including perhaps a solar power station. Such projects seem never to have obtained authorization in the past. In the early sixties, a proposal to build an experimental solar-tower power plant of 1,200 KW near Erevan was apparently seriously considered; some sources treat it as if it was actually scheduled to be built (see, for example, Zolotar'ev and Shteingauz, 1960, p. 178). But this proposal did not in fact get very far, and we can guess the reasons from an article in the Minenergo journal, which notes that it has very high capital cost per unit of capacity (*Elektricheskie stantsii* 1962:1, pp. 3–4). A similar project has now been suggested for the Turkmen SSR. An article in *Soviet News* (23 August 1977) says that such a station *will* be built, with a capacity of 100,000 KW, and that a new kind of design for the heliostats will cut the cost so much that cost per KW will be only half that of a conventional station. Such a claim suggests that this must be an early-stage proposal, and an interview with the President of the Turkmen Academy on the occasion of the creation of a new Solar Energy Institute in the Academy does not mention any plans for a solar tower (*Current Digest of the Soviet Press*, 28 March 1979, p. 24). It will not be surprising, however, to find the USSR eventually undertaking some analogue of the 10,000 KW installation that ERDA is now committed to building.

The implications of this brief survey of the solar energy program is that solar still seems very uncompetitive in the USSR, the planners see much cheaper alternatives, and in the absence of the kind of public enthusiasm and pressure felt in the United States, the Soviet effort has

*The NPO is a new kind of firm combining research and development units with production facilities intended to hasten the commercialization of new technologies.

been kept far smaller than the American, much more oriented to basic research, and still rather decentralized and unfocused.

CONCLUSIONS

This set of disparate examples suggests a few obvious points worth recapitulating. It certainly seems that the Soviet controllers do not hesitate to sponsor work on the speculative frontiers of energy R and D and are willing to support moderately expensive efforts in novel areas, even when it is difficult to make a case that these are likely to have large scale payoffs soon. The tidal, geothermal, and solar programs all support this conclusion. There also seems to be quite a bit of inertia in the system—R and D programs like those in coal conversion can drift for a long time, even if they do not seem to be closely related to high-priority tasks. But the institutional decoupling of R and D work from the major concerns of the powerful R and D sponsors that permit it to continue also means that it proceeds gropingly, without much focus or direction, underfinanced, and in a poor position to make a case for moving ahead, on the basis of findings at any one stage, to more ambitious tests.

It is clear that the system is not unwilling to give strong support to expensive and risky projects such as MHD or container pipelines. It is essential that an idea be taken up by some powerful sponsoring structure, in which case it can then survive all adversities. Unfortunately these cases do not reveal much about the process by which sponsors screen the menu of possible development ideas to determine those to which they will give their loyalties. Rather than try to exhaust the lessons implicit in the bases reviewed in this chapter, however, let us leave the subject here, to return to it in the final chapter, when an attempt will be made to draw together some generalizations based on all the chapters.

7   Technology Transfer

The purpose of this chapter is to examine some aspects of Soviet import of technology in the energy sector—such technology transfer being a form of innovation, an alternative to domestic R and D effort.

Soviet policy with respect to borrowing technology in the energy sector has been heterogeneous—there has been heavy reliance on borrowed technology in the oil and gas industry, but little use of this avenue for other branches. In the oil and gas industry, the USSR has imported line pipe, gas compressors, bits, a bit plant, pipe-laying equipment, exploration equipment, offshore drilling rigs and pipe-laying barges, gas processing equipment for the Orenburg field, and submersible electric pumps, to mention only some of the most prominent examples. In coal mining there have been only isolated cases of importing foreign technology embodied in equipment—the most notable instances being wheel excavators from Eastern and Western Europe, some transport equipment for open-pit mining from Eastern Europe, and machinery to equip the open-pit mine being opened at Neriungri to produce coking coal for export to Japan. In the electric power branch there seems to have been very little import of the basic equipment important in the branch; in fact, the USSR has been a successful competitor in exporting to world markets some of its domestically produced items such as hydraulic turbines, turbines and generators for thermal power stations, and in one instance a nuclear reactor and associated equipment for a power station.

It is interesting to speculate as to the causes of this difference in attitudes in the different sectors. Furthermore, experience in the two nonborrowing sectors has been very different—domestic development of technology in the electric power sector has been relatively effective, whereas reliance on indigenous R and D in the coal industry has led to a technological level far below that operating in other countries. We are interested in two main kinds of questions. (1) The borrowing of technology from abroad is an alternative to domestic innovation efforts, and we would like to find out what, if anything, we can say about how the decision to import technology is made in the USSR. The two approaches are not perfect substitutes, of course, since the absorption of foreign technology involves many of the same problems that absorption of a domestically created new technology does. Furthermore, depending on the form in which technology is transferred, imported technology may not provide the benefits of learning and building experience and capacity for further innovation. (2) A related question is how significant a contribution technological imports have made. This is an important issue in helping to understand the choices the Soviet planners have made about borrowing versus autonomous development, and this information is also relevant to business and government policymaking in the advanced market economies. Rather than try to survey or catalogue all these cases, it will be more useful to examine more carefully several specific instances of borrowing for which it is possible to get some quantitative idea of the payoff to borrowing. There are three examples we should look at in detail: in building its gas pipeline system the USSR has depended heavily on imports of large diameter pipe; more recently it has begun to import significant amounts of compressor equipment; to maintain output in depleted oil fields it has imported significant numbers of electric submersible pumps. These are sufficiently simple and distinct systems that it is possible to make some estimates of the payoff due to the imports. Since oil and gas are exportables with elastic demand, it is also relatively easy to figure the gain in terms of the net foreign exchange payoff.

GAS PIPELINES

Technical Level of Soviet Gas Transport

The technological inputs that govern the productivity of a gas pipeline system are numerous and varied, but two elements in the system are of central importance—the pipe itself (with its qualities of diameter, strength, integrity over time and under adverse conditions) and the compressors that maintain pressure in the line. There is a strong implication in the aggregate statistics of the Soviet gas pipeline network that its productivity, and hence presumably its technological level, is low. The transport work performed by the Soviet gas pipeline system and various elements determining the capacity of the system are shown in Tables 7-1 and 7-2. A comparison with similar data for a couple of years for the U.S. system, shown in Table 7-3, suggests that the work performed by the Soviet system is surprisingly small in relation to various capacity-determining dimensions.

TABLE 7-1. Gas Transported on Mingaz Lines

Shipments (10¹²m³)

	Total	Net deliveries	Losses and own use	Average distance shipped (km)	Transport performed (10¹²m³km)
1959	18.7	18.2	0.5	554	10.374
1960	26.8	26.0	0.8	589	15.782
1961	38.6	37.4	1.2	601	23.200
1962	52.3	50.8	1.5	611	31.950
1963	73.5	71.1	2.4	616	45.270
1964	90.1	87.5	2.6	644	58.024
1965	107.0	103.3	3.7	656	70.299
1966	124.6	119.8	4.8	678	84.450
1967	139.4	132.9	6.5	744	103.745
1968	151.7	145.7	6.0	864	131.168
1969	157.0	156.2	0.8	909	142.671
1970	179.1	170.1	9.0	917	164.237
1971	195.3	186.9	8.4	964	180.601
1972	205.6	195.6	10.0	1004	197.800
1973	217.8	205.5	12.3	1051	228.888
1974	243.4	226.2	17.2	1154	280.883
1975	272.7	250.6	22.1	1257	337.300
1976	302.2	279.5	22.7	1341	405.300

Shipments, Total: 1959–66—Gal'perin, 1968, p. 21; 1967–73—Khaskin, 1975, p. 15; 1974—by working back from net deliveries; 1975–76—*Ekonomika gazovoi promyshlennosti*, 1977:11, p. 24.

Shipments, Net Deliveries: 1959–66—Gal'perin, 1968, p. 21; 1967–73—Khaskin, 1975, p. 15; 1974—*Ekonomika gazovoi promyshlennosti*, 1977:5, p. 7; 1975–76—*Ekonomika gazovoi promyshlennosti*, 1977:11, p. 24.

Shipments, Losses and Own Use: by subtraction, except 1974, which is interpolated. There is clearly something fishy about the 1969 figure.

Average Distance Shipped: 1959—implied by data for work and shipments; 1960–73—Khaskin, 1975, p. 45; 1974—*Gazovaia promyshlennost'*, 1975:5, p. 8; 1975–76—*Ekonomika gazovoi promyshlennosti*, 1977:11, p. 24. These figures differ from others often quoted, which are figured by dividing work done by net deliveries rather than total shipments.

Transport Performed: 1959—Gal'perin, 1968, p. 31; 1969–73—Khaskin, 1975, p. 45; 1974—by multiplication from columns 1 and 4; 1975–76—*Ekonomika gazovoi promyshlennosti*, 1977:11, p. 24.

A pipeline system is made up of many segments of different sizes linked in various ways. To simplify our understanding of how much gas it can transport, however, we can think about it in the following way: in relation to the amount of gas that moves through the system (throughput), think of an *average* short segment, the most relevant characteristic of which is its cross section. As Table 7-3 shows, in 1970

TABLE 7.2. Capacity Indicators for Mingaz Lines

Year	Length of Lines (10^3km)		Average Diameter (mm)		Compressor Capacity (10^3kw)
	With Branches	Without Branches	With Branches	Without Branches	
1958	N.A.	N.A.	481		
1959	12.649	11.102	514		130.2
1960	16.573	14.392	553	683	256.7
1961	20.182	16.678	574	703	564.7
1962	24.441	19.746	581	710	910.2
1963	29.850	23.888	605	723	1190.0
1964	33.849	26.724	614	735	1538.0
1965	38.255	30.269	628	755	1868.8
1966	44.129	35.350	642	767	2069.0
1967	49.0			780	2460.5
1968	51.9			787	2990.7
1969	58.2			810	3077.3
1970	64.8		680	815	3400.7
1971	[68.5]			817	3873.7
1972	[74.7]			823	4348.3
1973	[80.5]			914	5309.8
1974	[89.1]			882	6712.0
1975	98.7		780	897	8232.1
1976	103.0			910	9603.1
1980	[135.0]				23,000.0

Length of Lines, With Branches: All data refer to end of year; 1959–66—Gal'perin, 1968, p. 7; 1967–70—TsSU, *Transport i sviaz'*; since 1970, data on the Mingaz share have not been published, but I adjust the figures given in TsSU, *Narodnoe khoziaistvo SSSR* for all pipelines by an estimated 3,000 km for local pipelines, based on their length in previous years; 1975, 1976—*Ekonomika gazovoi promyshlennosti*, 1977:11, p. 24; 1980 plan—*Gazovaia promyshlennost'*, 1977:11, p. 5.

Length of Lines, Without Branches: 1959–66—Gal'perin, 1968, p. 7.

Average Diameter, With Branches: 1959–66—Gal'perin, 1968, p. 33; 1970, 1975—Semenova, 1977, p. 10.

Average Diameter, Without Branches: 1960—Gal'perin, 1968, p. 33; 1961, 1965, 1970–73—Khaskin, 1975, p. 59; 1962–64, 1966–69, 1974—Furman, 1978, p. 72; 1975–76—*Ekonomika gazovoi promyshlennosti*, 1977:11, p. 26.

Compressor Capacity: (One kw = 1.36 HP.) 1959–60—Gal'perin, 1968, p. 34; 1961–73—Khaskin, 1975, p. 17; 1975–76—*Ekonomika gazovoi promyshlennosti*, 1977:11, p. 26. So far as I can tell, this is compressor stations on transmission lines, and I assume it is all on Mingaz lines. The figure for 1980 plan is from *Ekonomika gazovoi promyshlennosti*, 1977:11, p. 26.

the average cross section of Soviet lines was 1.79 times that of U.S. lines and, in 1975, 2.35 times as great, whereas Soviet throughput in these years was only 0.42 and 0.62, respectively, as great as U.S. throughput. Since for a given pressure, throughput should vary in pro-

TABLE 7-3. Comparison of U.S. and Soviet Gas Pipeline Systems

	USSR		U.S.		USSR/U.S.	
	1970	1975	1970	1975	1970	1975
Average diameter (mm)	680	780	508	508	——	——
Cross section (m²)	0.363	0.477	0.203	0.203	1.79	2.35
Length (10³km)	64.8	96.2	292.9	301.3	——	——
Volume of line (10⁶m³)	23.5	45.9	59.5	61.2	0.39	0.75
Compressor capacity (10⁶KW)	3.40	8.232	7.914	8.86	0.43	0.93
Throughput (10¹²m³)	170.1	250.6	486.2	402.1	0.35	0.62
Average length of haul (km)	917	1257	804.5	804.5	1.14	1.56
Transport work (10¹²m³km)	156	315	391	323	0.40	0.98

USSR: From Tables 1 and 2—refers to Mingaz lines and is based on gas delivered rather than shipped, as is true also for U.S. statistics.

U.S.: These data refer to the interstate network, with some small definitional departures. Compressor horsepower and length of system from FPC, *Statistics of Interstate Natural Gas Pipeline Companies*, 1970. Average diameter for 1970 is given in FPC, *National Gas Survey*, vol. III, 1973, pp. 126 and 23; and comparisons of the distributions by diameter for 1970 and 1975 in FPC, *Statistics of Interstate Natural Gas Pipelines*, suggest the structure did not change noticeably by 1975. Data on deliveries in FPC, *Statistics of Interstate Natural Gas Pipelines* is not quite suitable for our use, since it apparently includes a great deal of double counting. But a special FPC study, *National Gas Flow Patterns, 1975* (February 1977, p. 1), gives estimated throughput for 1975 on A & B interstate lines as the amount shown in the table, and we have estimated throughput in 1970 as the same ratio to gas sales by A & B interstate companies. There appears to be no standard statistical series on the average distance gas is transported on U.S. pipelines, but according to what I found in *The Economics of Soviet Oil and Gas* (p. 153), the average distance was about 500 miles in the sixties and is probably about the same now.

portion to cross section, the implication is that Soviet gas pipelines transport much less gas than they could be expected to.

There is a second dimension in the transport job in addition to moving the gas through short segments. The pressure drop due to friction as the gas moves through the line must be counteracted if throughput on an extended section is to be the same as through short sections, and this is the task of compressor stations. And the amount of this work will increase, the greater the average distance the gas must be moved. At one point the low throughput on Soviet lines was explainable by inadequate compressor capacity, but the data in Table 7-3 suggest that this is *not* the explanation now. As the table shows, by 1975 the Soviet network had more compressor horsepower per *volume* of line than did the U.S. network and about as much horsepower per cubic meter-kilometer of transport work. Both these measures reflect the distance dimension that compressors are dealing with. They thus allow for the fact that the Soviet gas must be transported on the average farther than the U.S. gas, and we are left with the fact that the considerably larger average diameter of the Soviet pipe ought to make it possible to move much larger amounts of gas through the Soviet system than through the U.S. system.

There are a number of possible explanations for this low productivity. First, the Soviet system probably uses somewhat lower pressures than do U.S. pipelines; most Soviet lines have been designed to operate at maximum pressures of 55 atmospheres (about 800 pounds per square inch). The first Soviet line designed to operate at 75 atmospheres (about 1,100 pounds per square inch) was commissioned only in 1972 (*Ekonomika gazovoi promyshlennosti*, 1977:5, p. 23). According to FPC, *Natural Gas Survey* (vol. I, 1973, p. 33), pressures in the range 700–1,000 pounds per square inch are common on U.S. pipelines, and some operate at pressures exceeding 1,000 pounds per square inch. This difference would mean that *design* capacities on Soviet lines may be somewhat smaller in cross section than U.S. lines. A second explanation is that even in relation to design capacities, the rate of utilization of Soviet gas pipelines is very low. One source gives the average percentage utilization of design capacity on the lines of the Ministry of the Gas Industry as follows:

1970—76.0 percent	1973—80.0 percent
1971—80.0 "	1974—82.0 "
1972—85.5 "	1975—80.0 "

[*Ekonomika gazovoi promyshlennosti*, 1977:5, p. 22]

The low utilization of rated capacity is the result of several factors. (1) There may be lengthy service interruptions because of breaks and leaks in the line pipe. (2) The rated capacity of the line may be seriously impaired because lines are badly fouled with condensate, water, and solids as a result of the failure to prepare gas adequately before shipment (*Gazovaia promyshlennost'*, 1978:10, p. 33). (3) The installed compressor capacity may not be fully available for work. Soviet compressor units apparently experience frequent breakdowns, and spend excessive time out of commission for repair or in reserve status. One author says that in the United States the planned availability of compressors is about 99 percent and actual availability 95–97 percent (*Gazovaia promyshlennost'*, 1975:9, p. 59). By contrast, the time budget for compressor capacity on Soviet lines in 1973 was as follows (in percent):

	gas turbine compressors	electric motor compressors	piston compressors
working	68.0	47.2	51.3
planned repair	8.2	3.0	9.0
forced outages	3.6	1.4	3.0
in reserve	16.3	21.4	13.5
no work available	3.9	27.0	23.2

Khaskin, 1975, p. 51

(4) There may be inadequate storage at the delivery end of the line to permit reasonably uniform utilization of capacity over the year. (5) It is quite possible that Soviet compressors are not efficient in terms of compression work performed per unit of capacity.

This analysis cannot be translated directly into a conclusion that the quality of Soviet pipe and compressors imposes a low technological level on the Soviet gas pipeline system. Through the latest year shown in the tables (1976), there was very little foreign compressor equipment in operation in the system, but the system has for a long time embodied extensive inputs in the form of imported large-diameter pipe. It does seem, however, that for the system the Soviet gas transport industry has built, it does not get the transport output it should. There is no doubt a large number of contributing factors, but there is a strong presumption that a low technological level both in terms of pipe quality and reliability of compressor stations is important in explaining the low productivity of the pipeline system.

Importance of Imported Pipe for the Gas Transport System.
A very large share of the increment in large-diameter pipelines since 1960 (defining large diameter as 1020mm and above) seems to have been accounted for by imported pipe, as suggested by the data in Table 7-4.

In 1961–1975, about 17 million tons of pipe were laid in oil and gas lines of 1020mm and above. Total imports of large-diameter welded pipe during the same period were 10.084 MT, or 58 percent. Moreover, dependence on imported pipe has increased over the period. In 1971–1975 total pipe investment in large-diameter lines was about 9.648 MT, of which 6.324 MT, or 65.5 percent, was imported. Those data refer to both oil and gas pipelines, but it seems likely that reliance on imported pipe is even greater for gas lines alone, since the quality demands for pipe used in gas lines are higher than for that used in oil lines.

Table 7-4. Gas and Oil Pipeline, 1020mm and Above

Year	Networth Length (10³km) Gas	Oil	Total Increment in Year Shown (10³ tons)	Imports of Large-Diameter Welded Pipe (10³ tons)
1960	0.670			266
1961	1.277			237
1962	2.020			523
1963	3.703			263
1964	5.104			166
1965	7.528	1.3		195
1966	10.112	1.3	904	175
1967	13.011	2.6	1,470	205
1968	14.054			436
1969	17.775			635
1970	19.132	3.9		925
1971	21.320	4.2	871	1,032
1972	25.336			994
1973	29.025			1,274
1974	34.738	11.0		1,332
1975	39.200	11.4	1,702	1,692
1976	42.024			
1980	[62.4]			

Figures in brackets are planned figures.

Network Length, Gas: 1960–66, 1969–Gal'perin, 1968, p. 33; 1967–68–*Gazovaia promyshlennost'*, 1970:4; 1970–75–Orudzhev, 1976, pp. 45–46; 1976–*Ekonomika gazovoi promyshlennosti*, 1977:11, p. 24; 1980 plan–*Oil and Gas Journal*, May 29, 1978, p. 99.

Network Length, Oil: 1965–67–Rubinov, 1977, p. 28; 1970, 1974–Dubinskii, 1977, p. 5; 1971–*Truboprovodnyi transport*, vol. 6, 1976, p. 13; 1975–Semenova, 1977, p. 10.

Total Increment in Year: Most Soviet pipe 1020mm and above has a wall thickness of 11–12mm, according to Spivakovskii (1967, pp. 68–69). The weight of such pipe is given by Friman (1976, p. 21) as follows: 1020mm–299 tons/km; 1220mm –358 tons/km; 1420mm–446 tons/km. We also know that for all gas pipelines laid in 1971–75, only 60 percent of which was 1020mm and above, the average metal investment was 334 tons/km. I assume that 350 tons/km cannot be far off for all pipeline 1020mm and above.

Imports of Large-diameter Welded Pipe: Soviet foreign trade handbooks. Unfortunately, beginning in 1976, large-diameter pipe was no longer shown separately. I have not found a source that could settle definitely what is included in the foreign trade commodity class "welded pipe of large diameter," but it seems very likely that imports were almost exclusively pipe of 1020mm and above.

There is a problem in reconciling these data with estimates of domestic production of pipe 1020mm and above, shown in Table 7-5. That table suggests domestic production of about 21 million tons of pipe 1020mm and above in 1961–1975, which together with imports gives a total supply of about 31 MT, whereas we figured the amount used in oil and gas lines as only 17.746 MT. There is a similar discrep-

TABLE 7-5. Estimate of Domestic Output of Pipe, 1020mm and Above

	All Pipe Output (MT)	Share of 1020mm and Above	
		Percent	MT
1960	5.805	1.93	112
1961	6.357	2.22	141
1962	6.878	4.03	278
1963	7.521	8.34	627
1964	8.124	11.7	953
1965	9.014	12.0	1,093
1966	9.905	12.2	1,210
1967	10.582	12.5	1,318
1968	11.215	13.0	1,453
1969	11.551	13.2	1,525
1970	12.434	13.7	1,701
1971	13.356	14.2	1,891
1972	13.829	14.7	2,031
1973	14.309	14.9	2,138
1974	14.961	15.1	2,262
1975	15.967	15.4	2,453
1960–1975	—	—	21,186
1976	16.806	15.6	2,622

All Pipe Output: This is a standard handbook series.

Share 1020mm and Above, Percent: This percentage is arrived at as the product of information on the shares of "large-diameter welded pipe" in all steel pipe output, and of the share of the class "1020mm and above" in all large-diameter welded pipe.

The former has changed little over this whole period—it was 22 percent in 1969, 23.5 percent in 1965, 24 percent in the early seventies (Poliak, 1965, p. 60; Belan, 1962, p. 187; Spivakovskii, 1975 pp. 59, 248, 279). The latter has risen rapidly from 8.78 percent in 1960 to 51.1 percent in 1965, 57 percent in 1970, and 61.2 percent in the early seventies (Spivakovskii, 1967, pp. 68–69; Spivakovskii, 1975, pp. 59, 248, 279).

I have interpolated between the given years and extrapolated to 1975 in each series and multiplied the results to get the series shown in column 2.

A possible source of error in these estimates is inconsistency in the definition of what constitutes "large diameter welded pipe"—in early sources it is described as 478mm and up, in later sources as 529mm and up, and in another case it is described as including 426–1640mm pipe (Tartakovskii, 1978, p. 5).

Share of 1020mm and Above: product of first two columns.

ancy for 1971–1975, with a total supply from domestic production and imports of 17.099 MT versus 9.648 MT accounted for by newly commissioned lines. Even if we assume an average lag of as much as three years between production and import on the one hand and commissioning of finished line on the other, there is a serious discrepancy between 12.623 MT supplied in 1968–1972, and the 9.648 MT in lines commissioned in 1971–1975. The gap might be the result of underestimating the average weight of pipe, a considerable share of sizes less than 1020mm in the import figure, or an overestimate of domestic production. Overall, however, I conclude that the ratio of imports to pipe in new lines figured above probably overestimates somewhat the Soviet dependence on foreign pipe.

Today the Soviet steel industry can itself produce and is producing 1020, 1220, and 1420mm pipe (*Sotsialisticheskaia Industriia*, 23 December 1975). One might therefore contend that the USSR imports pipe, not because of technological incompetence, but in order to evade production capacity bottlenecks or to avoid high costs at the margin from expanding domestic output. I would take the position that even these considerations reflect some technological weakness, especially in view of the fact that imports have continued over a couple of decades. In the years when pipe was imported, the Soviet authorities found it physically impossible, or at least excessively costly to organize the production of this pipe domestically, and that is more likely to represent a technical weakness than a temporary production bottleneck. Moreover, this is a technological import in the sense that Soviet pipe is of lower quality than can be obtained abroad, in yield strength, in wall thickness, and in general quality. So far as I can tell, Soviet pipe is inadequate for lines operating at more than 55 atmospheres. One author states rather diffidently that "we have experience with transporting gas at 75 atmospheres pressure in pipe of domestic manufacture (Tartakovskii, 1978, p. 6), but I believe that the newer lines operating at 75 atmospheres have been built essentially with imported pipe. The Russians produce both spiral-welded and straight-welded pipe, the latter from two strips so that there are two separate welds, which is said to add to the problem of inconsistency in quality (*Stroitel'stvo trubopro-vodov*, 1977:10, p. 31). So my case is that, if they had not been able to import pipe, they could not have created even the capacity the system presently has, because they could not have produced these sizes and strengths themselves.

Domestic Compressor Development versus Importation.

The USSR has had difficulty developing and producing compressor equipment domestically. A short historical sketch of the development efforts for turbine-powered compressor units shows frequent slippages, long delays in mastering the production of large-size units, and a frequent shift of tactics in response to technological failures.

Soviet pipeline designers have shown a general preference for gas turbine-powered centrifugal compressors as the basic equipment for large transmission lines. The rationale for this preference has been lower cost, autonomy from external electrical power supply, and the example of trends in other countries. As Table 7-6 shows, by 1976 the share of gas turbine prime movers in all compressor power on Soviet gas pipelines was 71 percent. Though this is a much higher share than in the United States, where it is more like 50 percent, it still represents a failure to produce the amount of this equipment considered desirable by Soviet gas industry planners. Electrically powered centrifugal- and piston-type compressors have been substituted to some extent to compensate for the failures and delays encountered in getting the desired gas turbine-powered equipment. As one author says in referring to the mid-sixties, "the high share of electric powered compressors resulted to a large degree because of the slow mastery of production of gas turbines" (Gal'perin, 1968, p. 36). The failure to produce the desired mix continued into the 8th Five Year Plan (1966–1970); it was hoped that gas turbines would account for 77 percent of capacity by the end of 1970, while in fact their share was only 57 percent (Gerasimov, 1969, p. 6).

The original Soviet gas turbine-powered compressor for gas pipelines, first produced in 1958, was based on the GT-700-4 turbine (capacity, 4 MW). That turbine, produced by the Nevskii machine building plant, was originally developed for driving power generators. The turbine was later adapted and upgraded to create more powerful gas pipeline compressor units with capacities of 4.25 MW, 4.4 MW, and 6 MW (Gerasimov, 1969, pp. 44, 139–40). This family of models became and remained the workhorses of the Soviet gas pipeline system right up to the mid-seventies.

Well before the seventies, however, an acute need for larger units was recognized. Between 1965 and the end of 1975, pipe with a diameter of 1020mm and above accounted for 55 percent of all gas pipeline added, and for these large diameter lines units much larger than 6 MW are desirable. Indeed a development program for 10 MW, 16 MW, and 25 MW compressor units had been instituted already in the sixties. But the program has progressed rather slowly, and its goals have been modified at various stages. The 10 MW unit was successfully put into production and has been widely used. During the 9th Five Year Plan (1971–1975) a large number of 10 MW units were installed, and by the end of 1974 they accounted for about 26 percent of all capacity, which implies about 174 units out of the 1,918 shown for that year in Table 7-6 (*Truboprovodnyi transport*, Vol. 7, p. 45). By the end of 1974, a prototype of the 16 MW model had been produced and accepted and recommended for production by an interdepartmental commission (*Gazovaia promyshlennost'*, 1975:1, p. 32). The assertion in the source that it would be used in compressor stations in 1975 was probably premature, though we know that three of them had been commissioned on the West Siberian lines by the end of 1976 (Bogopol'skaia, 1979, p. 5). There were also six of the GK-25I models installed on lines by the end of 1976 (ibid.). The program was amended in 1971 to include a 40 MW unit, but I have seen neither evidence nor claims that significant progress has been made on this model.

In 1974 it was decided to sidestep some of these obstacles encountered in developing and using the original family of models by moving to units based on aircraft engines (*Gazovaia promyshlennost'*, M, 1975, p. 68). One of the major attractions of the aviation-type units is their light weight, compactness, and block construction, which greatly simplifies and cheapens the construction of compressor stations. The Soviet approach has been to use a turboprop engine hooked directly to the compressor; this has made it possible to use the most powerful Soviet aircraft engine, the NK-12-ST, originally used to power the AN-22, the Bear bomber, and the TU-114. This, by the way, represents a case of reverse technology flow from military to civilian uses, a phenomenon that is often thought to be negligible in the Soviet economy. In its aircraft applications this engine was rated at 8.948 MW and then 11.033 MW in a redesigned version, but for compressor use it has been possible to get only 6.3 MW from it (*Janes' All the World's Aircraft*, 1977, 1978). This approach has also made it possible to sidestep the development of a new lightweight industrial gas turbine, a necessary element in the Western models, and a step that is difficult for the Soviet Union. This choice has, however, hindered achievement of the goal of easy installation and repair, since careful alignment of the turboprop engine with the compressor is not easy to achieve. Western models use a pure jet only as a combustion chamber so that there is no mechanical-drive connection between the jet engine and the turbine-compressor module. Also, the Soviet machine seems very cumbersome—its weight is 45 tons, whereas a 24.4 MW model produced by United Technologies weighs only 35 tons (*Gazovaia promyshlennost'*, 1978:8, p. 8; 1978:11, p. 40).

The prototype was accepted by Mingaz in 1972, and several such

TABLE 7-6. Compressor Equipment on Soviet Gas Pipelines (end of year, except for size of incremental units)

	Number of Stations	Aggregate Capacity Installed (10³ KW)	Turbine-Powered Centrifugal Compressors		Average Size of Unit (10³ KW)	Implied Number of Units	Size of Incremental Units (10³ KW)
			Percent	(10³ KW)			
1959	18	130.2	0	0	1.05	124	2.75
1960	21	256.7	0	0	1.51	170	3.46
1961	28	564.7	29	164	2.18	259	4.43
1962	37	910.2	26	237	2.70	337	2.69
1963	52	1,190.0	29	345	2.70	441	3.32
1964	71	1,538.8	30	462	2.82	546	3.32
1965	81	1,868.8	35	654	2.81	665	2.77
1966	85	2,069.0	40	828	2.98	694	6.90
1967	96	2,460.5	47	1,156	3.00	820	3.11
1968	119	2,990.7	50	1,495	3.07	974	3.44
1969	124	3,077.3	55	1,692	2.80	1,099	0.69
1970	130	3,400.7	57	1,938	2.92	1,165	4.90
1971	136	3,873.7	56.6	2,192	2.98	1,300	3.50
1972	150	4,346.5	57.5	2,499	3.07	1,416	4.08
1973	158	5,296.5	60.7	3,125	3.36	1,576	5.94
1974	173	6,712.0	66.2	4,443	3.50	1,918	4.14
1975	209	8,218.0	66.0	5,424	4.00	2,054	11.07
1976	214	9,603.1	71.0	6,818	4.17	2,303	5.56
1977	n.a.	11,200	n.a.	n.a.	n.a.	n.a.	n.a.
1978	n.a.	12,600	n.a.	n.a.	n.a.	n.a.	n.a.
1979	n.a.	(16,100)	n.a.	n.a.	n.a.	n.a.	n.a.

Number of Stations: 1959–64—Kortunov, 1967, p. 101; 1965–75—Orudzhev, 1976, p. 47; 1976—*Ekonomika gazovoi promyshlennosti*, 1977: 11, p. 26.

Aggregate Capacity Installed: 1959–76—Same sources as for Number of Stations; 1977–79—*Gazovaia promyshlennost'*, 1979:5, p. 5.

Turbine-Powered Centrifugal Compressors: 1959–63—Gal'perin, 1968, p. 36; 1964–74—Orudzhev, 1976, p. 47; 1975—*Gazovaia promyshlennost'*, 1977:4, p. 10, 1976—Furman, 1978, pp. 57, 72. Absolute amounts calculated from percentages given in sources cited.

Average Size of Unit: 1959—Gal'perin, 1968, p. 35; 1960–73—Khaskin, 1975, p. 17; 1974—*Planovoe khoziaistvo*, 1975:1, p. 22; 1975—*Gazovaia promyshlennost'*, 1975:12, p. 3; 1976—Furman, 1978, pp. 57, 72.

Implied Number of Units: Calculated.

Size of Incremental Units: Calculated.

units were tested in the Kupianskaia compressor station on the line from Shebelinka to Ostrogozhsk. On the basis of this experience it was decided to produce 46 units to be used in 12 compressor stations on the Orenburg-Kuibyshev-Center line, and the Nizhniaia-Tura-Center line. Some of these stations were reportedly commissioned in 1974 (*Gazovaia promyshlennost'*, 1975:5, p. 7). The series model of this aviation-type unit (the GPA-Ts-6.3) has been in production since 1974, and a statement in late 1977 reports that about 150 units had been produced, with a total of 453.6 MW (or about half the capacity that 150 units would aggregate) installed in compressor stations (*Gazovaia promyshlennost'*, 1977:11, p. 38). By the end of 1977, there were stations with a capacity of 605 MW in operation (*Gazovaia promyshlennost'*, 1978:8, p. 10).

The prototype used an engine retired from aviation service, and the series model apparently uses large amounts of parts from retired aircraft engines (*Gazovaia promyshlennost'*, 1978:8). The early versions of this domestically produced aviation model did not work well, but by 1978 it was being described as a great achievement. Apparently, however, the planners do not see this as a solution to the need for large amounts of compressor capacity, since the USSR has now ordered a significant number of aviation-type units abroad, units which have a capacity double that of the Soviet model (see below).

Another expedient announced as being tried is pipeline compressors powered by marine gas turbines with capacities of 10 MW and 16 MW (Orudzhev, 1976, pp. 65–67). Orudzhev has subsequently said that early experiments with these engines have been successful, and that more powerful models of these machines should be produced for the northern lines (*Gazovaia promyshlennost'*, 1978:4, p. 10). That sounds as if heavy use of these units is unlikely to come for several years.

Current development objectives include redesigning and repackaging units with capacities already mastered (i.e., the 4–6 MW models) into modular blocks. This program also has fallen far behind schedule. According to an early story seven such models were to be mastered in 1973–1974, but mastery of only three was actually begun (*Gazovaia promyshlennost'*, 1974:9, p. 12). A prototype (*golovnoi obrazets*) of the GTN-6 (produced by the Urals plant and the first model in this modular "third generation") was supposed to complete test trials and be approved by an interdepartmental commission in 1975. The goal was to produce it in 1975 to power stations on the Punga-Vuktyl-Ukhta lines (*Gazovaia promyshlennost'*, 1975:11; 1975:2; 1975:7), and it is reported that series production did begin in 1975 (*Gazovaia pro-myshlennost'*, 1977:11, p. 43). The prototypes (*golovnye obraztsy*) of the GTN-10, GTN-16, and GTN-25 compressors were retargeted for production by Mintiazhmash by 1976 (*Gazovaia promyshlennost'*, 1975:6, p. 6); by the end of 1976 a prototype of one of these had been produced—the GTN-16—which it was hoped might be in production by 1979 (*Gazovaia promyshlennost'*, 1977:11, p. 43).

The major models using an electric motor as prime mover (produced originally at the Nevskii plant but then at Khabarovsk) have capacities of 4 MW or 4.5 MW. The most common piston models have about 1,000 KW capacity (Gerasimov, 1969, pp. 134–137). Development of a 3.4 MW piston model was begun in the sixties, though in 1969 it was still not mastered (Gerasimov, 1969, p. 11). The largest piston unit attempted is one with 5,000 HP (or 3.7 MW) capacity, but it is not clear that it has ever emerged from the R and D process. In any case it is intended for storage reservoirs and head station use rather than for transmission use.

The best indication that none of these efforts to produce large-capacity models has been very successful are the data in Table 7-6 showing the average size of the incremental units in the stock. As the table shows, the average size of all compressor units in service on transmission lines has remained relatively small over the whole period, and even for compressors newly installed the average has moved up only very slowly.

The models that have been produced apparently require complicated and expensive installation work. In typical Soviet fashion they are not shipped as units and must be assembled on the construction site. This is a second source of delay in getting compressor capacity commissioned, and the plans for adding compressor stations have generally been badly underfulfilled. In the 9th Five Year Plan (1971–1975) the mileage targets for new pipeline were basically met, but the targets for compressor stations were met only in part, as shown by the following figures:

> 1971 57.6 percent
> 1972 40 percent
> 1973 52 percent
> 1974 57.1 percent
> 1975 56.0 percent

[*Ekonomika gazovoi promyshlennosti*, 1977:4, p. 9]

Domestically produced compressors also seem to have very unsatisfactory service lives. The GPA-Ts-6.3 is being touted as a great success for having achieved in 1978 an average time to breakdown of 1,970 hours (*Gazovaia promyshlennost'*, 1978:8, p. 19). This in spite of the fact that the guarantee period for this model was originally set at 4,000 hours between capital repairs and, from 1977, was raised to 8,000 hours (*Gazovaia promyshlennost'*, 1978:8, pp. 11, 26). American firms are advertising in the Soviet journals models, individual units of which have performed 25 to 40 thousand hours of continuous service (*Gazovaia promyshlennost'*, 1978:11, p. 41 and back cover).

The obstacles to meeting the need for compressors by domestic development seem to be the usual combination of purely technical difficulties (as in metallurgy), organizational and managerial weaknesses, and poor R and D management. Since the experience with aviation engines and with gas turbines for power generation has been similar, it is probably justifiable to conclude that gas turbine technology is simply a tough technology for the USSR to master. In any case it seems abundantly clear that the machinery plants just cannot produce the promised models on the requisite timetable, and it is easy to understand the Soviet decision to turn to import of foreign compressors on a large scale.

The import of gas line compressor equipment is difficult to follow in the official trade statistics, but it may be reconstructed from other sources. In 1973, the USSR made its first deal for importing large gas turbine-powered compressors and, by the end of 1976, had ordered about 3,000 MW of such units (*Gas Turbine World*, July, 1976). The main deals in this import program were as follows:

(1) In 1973, 37 Solar units of about 2.65 MW capacity were each ordered (and apparently delivered in 1974, to judge from the bulge in the trade statistics of that year), for an implied aggregate capacity of 98 MW. They were installed in the Ukraine, though I do not know on what line (*Gazovaia promyshlennost'*, 1975:3, p. 4; *Oil and Gas Journal*, 9 July 1973).

(2) For the Bratstvo line, 63 GE gas turbine units were ordered. These apparently had capacities of 14,600 HP (i.e., 10.9 MW) each (Moscow Narodny Bank, *Press Bulletin*, 8 August 1976, p. 1; and 15 December 1976, p. 3) and were produced by several companies. This lot of compressors would have a total capacity of 686.7 MW.

(3) For the Soiuz line from Orenburg to eastern Europe, 158 GE gas turbine units were bought, the production of which was divided among

several companies (*Wall Street Journal,* 4 June 1976; and Moscow Narodny Bank, *Press Bulletin,* 15 December 1976). There will be 22 stations on this line. I gather that those are similar to the units ordered
earlier for the Bratstvo line—i.e., with capacity of about 10.9 MW each
and implied aggregate capacity of 1,722.2 MW. There must have been
some other contracts as well, since it is said that "over 240" of these
GE units (compared to the 223 total mentioned here and in the paragraph above) have been bought (Moscow Narodny Bank, *Press Bulletin,* 24 August 1977).

(4) The USSR ordered 42 units using the Rolls Royce Avon engine
in combination with a lightweight industrial turbine for the Chelya-
binsk line (Moscow Narodny Bank, *Press Bulletin,* 15 December 1976).
The average capacity of these is apparently about 13.5 MW (*Gas Turbine International,* July/Aug. 1977, p. 29).

Overall, I read this record as clearly demonstrating an inability to
develop the gas turbine compressors, or at least an inability to produce satisfactory ones on time, and in the end a decision to resort to
foreign firms for compressor technology.

Benefits from Importing Pipe and Compressors

It would be gratifying to be able to conclude this section by quanti-
fying the net benefits generated for the Soviet economy, by pipe and
compressor imports, in somewhat the same way Philip Hanson calculated the benefits from importing fertilizer plants. There is too little
information to permit doing this seriously, but we can find some approximate data that give a notion of how beneficial these imports have
been. One might ask what the capital cost would have been to provide the same gas transport capacities with purely domestic equipment
rather than with imported pipe and compressor units. Planning handbooks give varied figures for construction costs of compressor stations
with domestic equipment, as for example the following:

> Using GKT-6—191 rubles/KW (Khaskin, 1975, p. 69)
> Using GTK-10—180 rubles/KW (Khaskin, 1975, p. 69)
> Using GTK-10—146 rubles/KW (Semenova, 1977, p. 77)

The Semenova book suggests that only about half the figure quoted
represents the equipment itself. But of course the Western equipment
saves much of the expensive construction and installation cost that is
needed for the Soviet equipment.

Another calculation shows the following: total investment in gas transport in the 9th Five Year Plan was 6.260 BR (*Ekonomika gazovoi promyshlennosti*, 1977:5, p. 5). According to Semenova (1977, pp. 70–71), compressor stations account for about 20 percent of the cost of large pipelines, which implies an investment in compressor stations of about 1.252 BR; if this is divided by the increment in compressor station capacity over the same period, average cost is 259 rubles per KW. The 3,074 MW of foreign compressor equipment listed earlier cost the Soviet Union about $1,199 million (same sources as quoted for the import deals). If we assume that the investment cost using domestic equipment would be somewhere between 150 rubles and 200 rubles per KW, the dollar/ruble ratio for this equipment is between 2.6 and 2.

For pipe, the imports of 10.048 million tons of pipe in 1961–1975 cost 2.325 billion foreign trade rubles. Over that period, the foreign trade prices have been converted to foreign trade rubles at varying exchange rates but mostly at about 1.33 dollars per ruble, which implies a 3.093 billion dollar expenditure of foreign exchange. The average wholesale price of the better qualities of domestic pipe, 1020mm and above, as of 1976 varied from 182 to 258 rubles per ton (Semenova, 1977, p. 84), and if we use a price of 200 rubles per ton, the cost of the pipe if domestically produced would have been 2.010 billion rubles, for a dollar/ruble ratio of just over 1.5. There do not seem to be any good studies of the domestic cost to earn a dollar's worth of foreign exchange, but Treml has estimated that in the sixties the ratio of foreign exchange ruble valuations to domestic valuations on all exports was about 1.21, when the official rate of exchange was 1.11 dollars per foreign-exchange ruble, implying that a ruble's worth of exports earned on the average 1.34 dollars worth of foreign exchange (1972, p. 163). On the basis of these figures it would be difficult to argue a huge resource saving from the imports.

There is, however, a second, independent gain in the form of accelerating the creation of gas transport capacity. It seems likely that, left to its own devices, the Soviet economy would not have been able to provide the transport capacity to get the gas to market without considerable delay. Suppose that there had been no imported equipment and that, as a result, the delivery trajectory realized would have lagged behind the actual by one year. To make it simple, assume that domestic outlays are the same for the two variants. That is, assume that the time profile of domestic resource outlays, using domestically produced pipe and compressors would be no different from that required to follow the import alternative—all that is different is the time profile of when they get this capacity in operation. This seems a minimal estimate of how great a delay would have been imposed by a lack of access to foreign equipment. The difference figured as of the mid-seventies would then be about 20 billion cubic meters of gas lost forever. That would be worth about 750 million dollars at the export price of the mid-seventies and would justify considerable expenditures on foreign equipment.

SUBMERSIBLE ELECTRIC PUMPS

A third notable instance in which the USSR has resorted heavily to imports of technology is that of electric submersible pumps. Such pumps are used primarily to enhance the flow from a well from which there is no longer sufficient reservoir energy to force oil at a satisfactory rate. Compared to other forms of mechanized lift (gas lift or pumps worked with pumping jacks), electric submersibles have higher outputs, do not require elaborate installations for above ground equipment, and should require less maintenance. The Russians have been using electric submersibles to some extent since the mid-fifties and, as old fields have become depleted, have come to resort to them more and more, especially in the Volga-Ural region. In West Siberia they are seen as an important aid in the effort to intensify production from fields, even at an early stage of development.

The first submersibles the Russians used—in the Ishimbai field in 1941–1943—were imported probably under Lend Lease (Amiian, 1962, p. 83). Until the 1970s, however, they subsequently depended on domestically produced pumps. Domestic production was deficient in several respects.

A relatively small selection of models was produced, and those that *were* produced tended to be of low capacity. Some 42 models are listed in Murav'ev (1971, pp. 190–91), including some up to 86 HP, but apparently only some of those listed were ever actually produced. In the Volga-Ural region on 1 January 1975, 81 percent of the pumps on hand had rated capacities of 100 m³/day or less (Galkin, 1966, pp. 114–115—these rated capacities are stated for work on water, and actual capacities for oil-water emulsions are much less). A. P. Krylov states that, at that time, there were only 24 models available (*Neftianoe khoziaistvo*, 1973:1, p. 23). Over the years there is a great deal of discussion to the effect that industry is "mastering" the more powerful models, that an experimental lot of some model has been sent for testing, and so on, with subsequent statements indicating that these plans never were carried out or were much delayed. For example, V. D. Shashin, Minister of the Oil Industry, was saying in 1968 that it was essential to get more powerful pumps that could be used in 6-inch casing, but only in 1973 were such pumps tested and series production begun (*Neftianoe khoziaistvo*, 1968:6, p. 5; and 1974:3, p. 7). The design characteristics of submersible pumps have to be adapted to the specific and highly varied conditions of individual wells, and the limited selection meant poor utilization of the capacity of those the Russians did produce. The pumps finally bought from the United States were rated at 165–215 HP (*Oil and Gas Journal*, 29 January 1973) and had capacities of 800–1,000 cubic meters per day (Muravlenko, 1977, p. 83); apparently it was only in 1975 or 1976 that domestically produced prototypes of similar capacities had reached the test stage.

Moreover, the Soviet equipment has been low in quality and reliability, so that periods of operation between breakdowns and overall service lives are very short. There is a lot of contradictory evidence on the actual between-repair periods. What purports to be a statement for the USSR as a whole puts the between-repair period for 1975 at 225 days (*Neftianoe khoziaistvo*, 1974:7, p. 29), though many other accounts suggest much shorter periods. U.S. pumps are said to be designed to operate for 500 days continuously (A. A. Meyerhoff). Quality deficiencies are made worse by the common Soviet weaknesses of limited selection and unavailability of repair parts, so that, in the field, producers are often forced to perform makeshift repairs and content themselves with holding a very large share of the available pumps idle as a reserve. One source complains about the high incidence of pumps which, on being started up again after repair and reinstallation, fail immediately (*Neftianoe khoziaistvo*, 1975:6, p. 33).

As the Volga-Ural fields became more heavily invaded with water, it became necessary to lift very large volumes of fluid to maintain oil output, and the need for high-capacity pumps became acute. The Soviet oil-field equipment industry has long had programs for developing pumps with capacities of 1,000 cubic meters per day, but it was not until 1975 that domestic R and D had produced some models of this capacity for testing (*Neftianoe khoziaistvo*, 1975:3, p. 24). Given the urgency of maintaining output from declining fields, the oil industry could not wait, and in the 1970s it was decided to acquire such pumps from U.S. companies. During 1972–1977 the USSR imported about 1,020 of these pumps, and there are additional orders outstanding for approximately 300 more to be delivered in 1978–1981.* The aggregate lifting capacity of the 1,020 pumps already delivered is about 3.5 million barrels per day, for an average of 546 cubic meters per day, though some of these pumps have capacities up to 1,000 cubic meters per day.

It seems likely that the availability of the U.S. pumps has been crucial in achieving output goals during the seventies, though there are many uncertainties in estimating how much of a boost these imported pumps have given to output. Meyerhoff says that, because the Soviet clients did not give the American suppliers sufficient freedom in adapting these pumps to the specific locations and because they have not used the U.S. companies' help in maintaining them, the pumps failed to achieve anything like their potential productivity. Also, it is reported that many of these pumps have already been withdrawn from service. Nevertheless, we can probably attribute millions of tons of extra output to this import decision, along the lines of the following calculations.

It seems clear that the Soviet electric submersibles have a much lower productivity than the American pumps. The first two columns in Table 7-7 summarize Soviet statements as to the number of wells

TABLE 7-7. Electric Submersible Pumps in Soviet Oil Production

Year	Number of Wells Equipped With Submersibles (end of year)	Output From Those Wells (MT)	Oil Output/ Well/Day (m³)
1959	838	n.a.	—
1960	1,000	9.0	31
1964	2,665	29.2	37
1967	3,500	57.9	57
1969	5,200	n.a.	—
1972	6,285	115.0	66
1974	8,683	151.0	60
1975	8,585	163	65

Number of Wells Equipped with Submersibles: 1959, 1972—*Umanskii*, 1974, p. 101; 1960—Umanskii, 1962, p. 127; 1964, 1974—*Ekonomika neftianoi promyshlennosti*, 1975:7, p. 7; 1967—*Neftianik*, 1968:9, p. 6; 1969—*Neftianoe khoziaistvo*, 1972:1, p. 53; 1975—*Ekonomika neftianoi promyshlennosti*, 1977:11, p. 38.

Output from Those Wells: 1960, 1964, 1972—Campbell, 1976, pp. 27–29, 102; 1967 —*Neftianik*, 1968:9, p. 6; 1974—*Neftianoe khoziaistvo*, 1975:6, p. 33.

equipped with such pumps and the output produced from those wells. Dividing output each year by the number of wells (pumps) working

*Another source says that the USSR has imported 1,200 of them (Meyerhoff).

at the end of the year and assuming a specific gravity of 0.8 for oil, gives an approximation of pump productivity in terms of oil shown in the third column (I have converted from annual to daily output by dividing by 365, though obviously these wells worked less than 365 days a year because of breakdowns). The oil/liquid ratio in the output of wells produced with submersibles was 39.8 percent in 1960 (Amiian, 1962, p. 89), making productivity in terms of *liquid* 78 cubic meters per pump per day in 1960. The oil/liquid ratio has surely fallen significantly since then, and if it was as low as 25 percent in 1972, average productivity would be 264 cubic meters of liquid per day. In 1972 there were still no U.S. pumps, so we might take that figure as an indication of the productivity achieved with Soviet pumps.

If we assume that by 1975 as many as 600 of the American pumps were in operation, with an average productivity of 500 cubic meters of liquid per day, and that the other 7,985 pumps (Soviet) had an average productivity of 250 cubic meters per day the liquid raised in 1975 would have been:

by U.S. pumps	110 million m³
by Soviet pumps	725 million m³
Total	835 million m³

Given that *oil* output from wells equipped with electric submersibles was about 165 MT (206 Mm³ at specific gravity of 0.8), this would make the oil/liquid ratio about 0.25. If we figure the contribution of the American pumps on the basis of the differential average productivity and assume no difference in the oil/liquid ratio, the availability of the American pumps gave the Russians an extra 13.5 MT of oil above what they could have produced without these pumps. That is probably an exaggeration, since the American pumps have probably been most heavily used in wells with the lowest oil/liquid ratios. But, however much we play with these figures, that extra output attributable to being able to use the American technology is very large. Moreover, it seems to have been an excellent bargain for the USSR. The Soviet foreign trade handbooks show an average price on oil exports to hard currency countries in 1975 as 60.4 foreign trade rubles/ton, f.o.b. the Soviet border. During 1975 the prices at which exports were sold to hard currency areas were converted to foreign trade rubles at 1.32 dollars per ruble, making the average price earned on oil exports 79.7 dollars per ton. At that price, 13.5 MT is worth 1.1 billion dollars. Considering that the pumps cost probably no more than 110–120 million dollars, they seem to have a very high benefit cost ratio. (An article in *Oil and Gas Journal*, 29 January 1973, says a contract for 100 of the pumps was worth 6 million dollars; another source quotes 110–120 million dollars for the total including spare parts.)

DECISION-MAKING CRITERIA AND CALCULATIONS FOR TECHNOLOGY IMPORTS

The next task is to explain how these technology import decisions are made. The logical way to think about this is to suppose that planners set up some calculation in which the costs and benefits of import versus domestic development would be assessed in the usual cost-benefit manner. There is an interesting discussion in *EKO* (1972:4) that illustrates what some of the variants in such a decision might look like. That discussion puts heavy emphasis on the long-range gains, difficult to quantify, from developing domestic technical competence. The big trade-off is cost savings by import versus building scientific and technical capabilities. I believe, however, that this is not a good representation of what really happens; the more carefully I examine these three examples the less it seems that the USSR has any strongly institutionalized way of making a careful choice between domestic R and D and foreign technology.

There is quite a bit of evidence as to how the gas industry planners think about the value of pipe of various technical characteristics. The general approach is to start with some value for gas delivered at the major consumption areas, the production cost of gas in West Siberia, and the cost of various other inputs, using these to impute a value for pipe, which can be thought of as the limit price for pipe, above which building the pipeline and shipping the gas is not economically justified. Within that limit, the cost of alternative designs for the pipeline is compared according to the criterion of minimizing *privedennye zatraty* (i.e., operating cost plus an opportunity cost charge for the capital committed). This method is well illustrated in Tartakovskii (1978). He uses the method in a quite sophisticated way to evaluate the effectiveness of employing alternative grades of steel that would permit changes in line pressure, wall thickness, diameter, and so on. It is interesting, however, that he never considers the alternative of *imported pipe* for its potential to improve the "economic effectiveness" of the pipeline. One of the most interesting features in the book, however, is a conclusion that, with gas worth 30.8 rubles per 1000 cubic meters in the consuming area, the limit price for pipe is 913 rubles per ton. Since the value of gas must be still higher as a replacement to free oil for export or as an export itself, the limit price for pipe intended to support exports must be far above that figure. Since we earlier figured the import cost of pipe at $308 per ton, it would seem that, if they are simply unable to produce the needed pipe domestically, they could afford an import price far above what they actually pay.

Another author does take this line of reasoning further to evaluate the desirability of importing pipe. He seems to be saying that the "metal balance" is strained (i.e., that the domestic supply of pipe available for gas transport is fixed) and that in this situation it is appropriate to start with the shadow price of gas and work backwards to impute a worth to imported pipe. If pipe can be purchased abroad at a price lower than that, the pipe should be imported (Ushakov, 1972, p. 110). The author seems to envisage using some shadow exchange rate for converting the cost of pipe in foreign exchange into domestic prices. There are many references in these discussions to such a rate under the name of the "import equivalent," which is usually described as the ratio of domestic prices (costs) of importables to their price in foreign exchange. I don't know how far the concept of the import equivalent has been officially legitimated and embodied in routine decision making, but it seems fairly common in the economic literature. Unfortunately, none of these discussions give much hint as to what the authors see as the actual import equivalent coefficient.

Another author has essentially the same idea, except that he interprets the problem with domestic provision of pipe as one of poor assortment and low quality. The first constrains optimal design, and the second means that optimal design, under the constraints of domestic pipe characteristics, will require lower pressures and thicker walls, than under those associated with imported pipe, and hence investments of a very large quantity of metal to do a given job. Use of the same tonnage of metal in the form of imported pipe will permit a larger delivery of gas, which can be evaluated at its shadow price (Smirnov, 1975, pp. 62–63).

One of the most elaborate and revealing of these discussions of the import decision, and one that introduces some new wrinkles, is an explanation by two economists from the All-Union Scientific Research Institute for the Construction of Pipelines of how they propose to choose the optimal technology import for compensation deals in gas (Vainshtein and Takhenko, 1975, pp. 142–180). The essence of their approach is to create a considerable number of alternatives, embodying variations in most of the conditions of the project. The variations that interest me here are those involving differences in the mix between inputs of Soviet equipment and foreign equipment. In each variant, imports and interest on the credit are paid off by exports, and the goal is to choose the variant for which the domestic gas availability stream compared to the domestic resource input stream is the most advantageous. In this approach, since the variants involve different degrees of dependence on foreign technology, the choice between technology transfer and domestic R and D falls out of the overall optimization.

The fascinating and ironic thing about each of these is that they settle the issue of technology import versus domestic R and D without ever confronting the choice explicitly. In the Ushakov and Smirnov examples the possibility of domestic R and D to improve pipe quality is simply ignored; in the last example, what is considered is not the cost of improving domestic technology but the cost disadvantage of using existing domestic technology.

I believe that the decision to import pipe has indeed been made in something like this fashion. The reason the pipe case does not look like an R and D issue is that it involves a technology that is firmly embodied in a tradeable product and in which the task of substituting domestic pipe for imports is more one of organization, calculation, and flexible adaptation than it is of drawing on the capacities of the R and D organization of the system. Solving the technological problem domestically is more a battle against the uncertainties of organization than against the uncertainties of the physical world. It does not seem to the policymakers that they can solve the problem of getting large-diameter pipe of the desired quality and assortment, by the time they need it, by expending R and D resources proper, so they do not think explicitly in terms of domestic R and D versus technology transfer.

We might say that they do see the trade-off as one of domestic innovation (by which I mean something broader than R and D proper) versus technology imports, but it becomes very difficult to get any quantitative expression of the cost of domestic innovation in this sense to weigh against the cost of technology imports. They are more likely to go with a basic judgement as to the feasibility of the domestic innovation alternative within a given time horizon.

The interpretation of the compressor case is probably similar. The

USSR has had an ambitious and long-standing R and D program for developing this equipment. Compared to pipe, compressors present a more typical R and D problem, in that domestic supply will involve creating new models that require extensive R and D. Some of the best equipped, generally successful, competent Soviet R and D facilities and resources have been put into this area—i.e., the resources of the Nevskii plant, LMZ, the Ural turbomotor plant, and the associated research institutes and design bureaus. Probably because it was thought that this effort could match the achievement of foreign companies, the planners did not at first consider importing the equipment.

Soviet industry has simply been unable to meet the timetable for producing the large compressors needed on the large-diameter lines, and the planners have perforce turned to foreign suppliers. A Soviet analysis of the optimal design and equipment for the Cheliabinsk line, for which the British aircraft-derived units have been bought, originally assumed it would use Soviet compressors—specifically the GTN-16 or GTN-25. But progress on these was apparently too minimal for that intention to be carried out (*Gazovaia promyshlennost'*, 1977:1, p. 9). Furthermore, it seems that only late in the game were GE units decided on for the Orenburg line. An interview with the head of the prime contractor in 1975 indicated that 25 MW units would be used—presumably the still to be developed GTN-25 (*Soviet News*, 10 June 1975).*

It is revealing that in the area of gas turbine technology the USSR originally tried to organize technology transfer on a cooperative basis, in areas nearer the R and D frontier rather than as a straight import of technology embodied in equipment. The State Committee on Science and Technology (GKNT) has a scientific technical agreement with General Electric, signed early in 1973, in which gas turbine research is one of the high priority areas (*New York Times*, 13 January 1973). The Soviet planners thought they had the basic capabilities for development work on this kind of technology well in hand so that they could produce their own equipment. Any borrowing would involve ideas and basic research and would be paid for by a reciprocal exchange in kind rather than by conventional exports.

The submersible pumps, also, seem to fit fairly well into this basic

*There is a similar example in the coal industry. It was originally intended to equip the Neriungri mine with domestic shovels (EKG-20), draglines (ESh-40/85 and ESh-65/85), and a domestic 180-ton truck (*Ugol'*, 1976: 2, p. 40). In the end, however, imported trucks and shovels have been used, no doubt because there was no prospect of getting the projected domestic models developed in time.

interpretive framework. The USSR has produced these pumps for a long time and in an important sense had command of the technology and some experience with it. They failed, however, to upgrade them in the relevant dimensions fast enough to meet urgent needs. It is interesting that the imported technology is taken as a kind of standard for which the domestic R and D effort should aim. We might think of their decision-making process not as one of deciding whether a given result is best achieved by transfer or domestic effort, but of accepting technology transfer only in extremis and then gauging the amount of resources they put into R and D as whatever is required to match that result domestically and to free them, as time passes, from import dependence.

These examples confirm some well-established generalizations about Soviet R and D. The Soviet R and D establishment has excellent scientific capability, but it does not extend down through the engineering and innovative stages, so that what I assume are good enough basic research, concepts, and design ideas fail to be embodied in reliable, series-produced equipment. But planning goes on in a context of taking at face value the promise of the R and D performers that they will be able to handle the assigned task. Only later is it discovered that failure has occurred at the production end of the "science-production cycle." At that point, the decision makers have a variety of choices: they can use second-best domestic technology (electrically driven centrifugal compressors, metal-intensive pipe, etc.); accept delays (not build compressor stations on schedule); accept the actual technological level achieved and design around it (install large reserve capacity to make up for lack of reliability in the equipment). But in the oil and gas ministries, at least, and in the era of detente it surely occurs to the higher levels fairly routinely that technology importation is another alternative. And it must happen more often now that, when the planners cost out this alternative, it is sufficiently attractive in relation to any realistic assessment of what their own R and D people will deliver that it could be a permanent solution. I imagine that the question always involves, as a separate issue, whether the required foreign exchange is available. It is not enough to demonstrate its economic advantage—that is only the first step. Surely one reason the oil and gas people have been able to persuade the decision makers to accept technology imports is the fact that these commodities themselves earn so much of the foreign exchange that is required; the easiest case of all must be when the project itself involves direct export earnings. Availability of foreign exchange becomes very important when one moves from an enclave project to something like the compressors imported for the Cheliabinsk lines, where the project is intended to serve domestic needs. In this case, the availability of the needed foreign exchange probably becomes an independent question, even if the project is demonstrated to be cost-effective by the formal calculation.

One can understand the reason for this; it is difficult to relinquish decisions about foreign exchange expenditures to local project makers or to put considerations like stimulus to domestic technical level into value terms. Even when economists suggest how the latter might be done, political decision makers prefer not to relinquish these decisions to economists or to whiz kids with models.

Another factor that undercuts any belief that the domestic R and D versus TT decision is based on the relative costs is that it is unrealistic to imagine that the USSR will ever really abandon an entrenched domestic R and D effort in any significant field. Even when technology is imported to meet a crisis, the domestic R and D program will continue but perhaps be retargeted to copying or bypassing the technology embodied in the imported innovations.

Some implications of this interpretation are worth spelling out. (1) Such technological borrowing can decrease the technological gap and save large amounts of resources used directly in production. It will not, however, save R and D resources for the USSR. (2) The biggest gains are likely to come from borrowing strongly embodied technology. I do not see that the USSR will get much out of the high-level scientific and technical agreements—in this kind of activity they are close to us in relative standing or can get what they need via other channels. In the area of fusion, for example, the USSR may be giving us as much as it is receiving. (3) The real problem in the Soviet system is in the organization and management of innovation. In this downstream end of the R and D spectrum we find feeling out, adapting, optimizing, with increasing knowledge and heavier commitments taken and a smaller range of choices and unknowns to play with. I do not see that they can do much in the way of borrowing in this range of the R and D spectrum, since the structure they work in just puts very stringent limitations on change. For instance, it was said that the transition to series-production on their aviation-type compressor was held up *for a whole year* because there was no testing stand for it. That is a goof so tied up in systemic cumbersomeness that no amount of outside help or advice could do much to correct it.

8   Conclusions

Let us begin this concluding chapter by reviewing some commonplace but fundamental ideas about R and D. As a framework for organizing the evidence presented in the previous chapters into conclusions about the peculiarities and effectiveness of Soviet energy R and D, it will be helpful to restate briefly what R and D is and what some of the major issues in R and D management are.

Research and development is a process of generating new knowledge and applying it to solve production problems. As this definition suggests, the process encompasses a number of quite different activities, from basic research to the commercial introduction of developed technologies, commonly thought of as lying along a spectrum of decreasing uncertainty, greater applicability to production concerns, and increasingly heavy resource commitments. The search for new knowledge that constitutes one pole of the spectrum must proceed in a way that is not specifically focussed on the task of solving a recognized production problem. At the other, actual introduction of new technologies that raise productivity must take as their point of departure, capabilities and potentials that are known with a considerable degree of assurance. In between is a great variety of studies, experiments, and evaluations, characterized by intermediate degrees of relevance, uncertainty, and expense. One of the universal tasks in any R and D enterprise is the balancing of effort among the different activities along this spectrum—between long-range speculative efforts intended to create quite novel solutions versus efforts to make more modest advances on the basis of fairly secure expectations of what will work. In bulk power transmission, to take a concrete example, there is a question of whether to seek increases in the amounts and distances involved, through small incremental changes in the parameters of a familiar approach or to abandon that as inadequate to the task and explore instead such novel concepts as cryogenic conductors or wave guides.

This balancing is not, however, primarily a problem of choosing one or the other option or even of static allocation between the various parts of the spectrum. Rather, the division of efforts along the spectrum at any given time should grow out of a dynamic process of deciding when it is time in any given area to advance on the basis of the new knowledge built up by speculative explorations to a narrower and more focused exploration of possible solutions to some problem. When has enough knowledge been gained about the behavior of a certain kind of coal to justify taking the concept of energy-technological processing from laboratory experiment to a pilot installation, from a pilot installation to an experimental plant, and from an experimental plant to a demonstration or commercial facility? When has the environment surrounding a problem so changed as to make a research idea or direction irrelevant, so that it should be dropped and the effort shifted to an alternative tactic or back to rethinking the fundamentals of the phenomena involved? When has enough experience been acquired so that it is possible to decide on a big resource commitment with a very high probability that the specific solution proposed can be made to work in a situation that will now involve the acquiescence and cooperation of a much wider circle of actors?

At this commercialization stage someone will have to design and produce the associated equipment to meet performance specifications, while others will have to be willing to accept it, fit it into their environment of production goals and conditions, and make it work in practice. Success here is a function not only of how well all the interrelated tasks at this stage are carried out, but also of how well previous steps have set the stage for this final commitment. For example, the preparatory decisions must not have mistakenly neglected better alternatives that will subsequently emerge to undercut the effort put into this one—if fission reactors are more advantageous than breeders, even the most dogged attempts to commercialize the breeder will be a fiasco. The developers must have performed enough work on components, materials, and production processes to assure that the new equipment, facility, or system can be made to work; they must have adapted the final outlines of the technology well enough to the conditions that will be faced so that it will be accepted as economic. Only then can it succeed in being widely diffused throughout the area of application.

Against the background of this description of the process of R and D, what is distinctive about Soviet R and D experiences and approaches? What kind of systemic peculiarities seem to account for the results we see?

I find it difficult to summarize the energy sector R and D activity we have described in the form of generalizations that hold up in all cases, but there does seem to be a number of distinctive features common to most of the examples we have looked at. At the same time there are important differences from the energy R and D experience of other countries, and one of the goals of this final summary should be to reflect not only on the way Soviet system characteristics make R and D outcomes different from what we see elsewhere, but also to seek the variables that condition differences of outcome among cases *within* the Soviet system itself.

First it is perhaps obvious, but worth repeating, that in many of the areas we have examined, Soviet technology planners were following development paths already well explored by other countries. In energy as well as in technology generally, the Soviet Union has been and, in many areas is still, on a lower technological level than the advanced market economies. The experience of other countries thus offers helpful guidance in deciding the best directions to follow in technological advance and in dealing with the basic difficulty in all innovation—eliminating uncertainty as to whether some idea can be made to work. The shift to narrow-web combines, the scaling up of excavators, use of trucks in strip mining and continued increase in their capacities, the use of gas turbine equipment for peak power generation are only a small sample of cases where a technological line had been chosen, developed, and demonstrated to be effective in practice elsewhere well in advance of Soviet efforts. The Soviet planners could both take that experience as indicating the correctness of the line and operate with the assurance that the new technology could be made to work.

The fact that in many cases R and D programs represented domestic catch-up along well-established paths has had an important impact on the R and D process by encouraging the location of innovation decision making at a high level in the hierarchical structure. There was a presumption that experts at the top, possessed of a special vision, could best discern the key technological trends in any area and choose the most appropriate ones to follow. This kind of centralization gives a strong strategic flavor to the whole process, which encourages both entrenched biases at the center that may be difficult to overturn through experience at the bottom and selective attention that may leave important tasks neglected. Recall for example the negligence toward developing peaking equipment compared to the tremendous stress laid on mastering large-size generating units and stations.

This study has been interesting to me in showing that in many cases the technological inspiration has come from countries other than the United States. Long-wall mining techniques seem to owe nothing to the U.S. example; the experiment with tidal power has drawn on French experience and even French equipment; the Soviet strategies of nuclear power development have rejected the U.S. line on breeders and have accepted the French and British view that they are feasible and economically attractive. There are even a few cases in which the source of the inspiration was Eastern Europe, notably in the case of rotary excavators. It is a little surprising to me that there has not been more borrowing from Eastern Europe. I would think, for instance, that Polish coal-mining technology would offer a number of examples of technological improvements the Soviet Union could emulate, especially since Polish mines work in a similar system setting.

When technological progress takes place in this manner, one of the corollaries is that R and D is less a matter of picking a way through the unknown than of choosing from a menu of demonstrated possibilities. The problem is to determine which lines of advance already pioneered elsewhere are economically appropriate under Soviet conditions and to assess how ambitious an advance domestic R and D performers can be expected to manage. I don't believe this study has shown that there is a uniform conclusion as to how well this job is done. On the whole it is my impression that Soviet energy technology planners do have a healthy sense of the peculiarities of their environment and are reasonably cautious in deciding what new ideas are feasible for them to try to develop domestically. The great emphasis put on the heat rate as a design objective for power generating equipment is a natural response to the relatively high cost of fuel in the USSR. I have been impressed with what seem to be rather reasonable approaches to the economic analysis of alternative technologies in electric power generation, with the ideas underlying the fuel optimization models, and with the approach to the economic evaluation of nuclear power. Using larger pipe diameters to compensate for deficiencies in attainable working pressure and compressor equipment should probably be judged a reasonable accommodation to the difficulties the pipeline designers see in trying to match Western steel and compressor qualities.

We have seen a few cases in which decision makers were overoptimistic about the ability of domestic innovators to solve some problem, resulting in the failure of R and D programs to achieve their objective. This would be my interpretation of the problem with gas turbines, for example. Very often, however, this kind of overoptimism is not fatal. There is often some way to deal with failures to meet technological objectives, as when the electric power industry compensated for the lower quality of steel by accepting operating temperatures for power blocks below those the engineers thought they had designed for.

Despite this generally emulative approach in energy R and D, however, we have shown that the system is perfectly capable of supporting work in novel areas and at the frontier of technological advance, even when the direction chosen has not been validated by foreign experience. Ideas and possibilities that are a long way from realization because they involve basic research or because their potential must be evaluated through a long period of experimentation *do* get supported. The Soviet energy "establishment" itself has been thoroughly unenthusiastic about either solar or geothermal power, but research programs in those areas have gone on through the influence of alternative sponsors such as the Academies of Science. Even organizations that might be expected to be biased toward current production tasks have often supported long-shot ideas—Minenergo has been willing to spend resources on tidal power and to risk prestige and quite large resources on MHD. It is interesting to try to understand how such work is protected from the short time horizon that indeed dominates most decision making in the USSR, as well as from the pressures to focus on solving immediate production problems. One explanation is that, despite operating with a generally mission-oriented perspective, R and D planners understand well enough that there should be a distinct mission conceived of as building up a scientific-technical backlog of ideas that might be useful at some future time—the so-called *nauchno-tekhnicheskii zadel*. This is reinforced in execution by a kind of institutional decoupling—the designers of the R and D system really do try to set up organizations that will be independent of the production people with their focus on current problems because they have high-level protection and independent financing. This is the positive side of the phenomenon noted in Chapter 2—the inertia and insensitivity of many R and D organizations that comes from assigned specializations and the prevalence of institutional, rather than project, financing.

But I believe we must also accept the idea that the high-level decision makers in the production establishment, who in the Soviet setting do play a dominant role in the system, are capable of vision, and are willing to accept significant risks in supporting R and D. There are examples of research directions which have had continued and substantial support even though they may be at variance with the general directions of world technological development or have been consciously rejected by decision makers in other countries. Support for the MHD program has continued through the ups and downs this idea has suffered in other countries. Similarly, within an example not discussed in this book—the turbodrill—the oil industry supported through thick and thin a completely distinct line for improving drilling technology (see Campbell, 1968). In both these cases I believe the explanation must be partly in terms of personal and bureaucratic influence. When a bureaucratic structure or a powerful individual is really convinced and committed to some direction, it will be doggedly pursued despite discouragements or the contrary example of other countries. As another example of independent innovation, the Russians have committed themselves to cogeneration essentially on the basis of the energy conservation argument, even though this was neither strongly born out by their own experience or validated by the practice of other countries. In this case, the explanation is partly a difference in the institutional setting—central planning means that the heat and power markets were subject to unified control, and the technological decision makers had the sense to exploit the novel opportunity thus presented.

I would like to suggest that the various devices that protect basic and speculative research in the system exact a serious penalty by complicating the problem of moving smoothly from one stage of the science-production cycle to the next. This is one of the most serious weaknesses of Soviet R and D and will be discussed more fully below. Here I want only to note that the institutional decoupling that protects speculative and basic research from the attentions of officials with short time horizons and production orientations also makes it difficult to accomplish a smooth transition from one stage of research to the next. Everything is either rather speculative research or overhurried movement toward industrial application. We saw this in the case of MHD, in the development of big power blocks, and in the nuclear program; I will expand on this "commercialization" problem below.

This study of energy experience suggests that Soviet R and D institutions probably fail to generate at each stage enough alternatives for evaluation, a finding that corroborates a widely-accepted generalization in the Western literature on Soviet R and D. In Chapter 6, it was shown that importation of technology as an alternative to domestic innovation seems not to be a routine option in the economic analysis of R and D decisions. There is a compartmentalization of the decision-making environment which requires some extraeconomic bureaucratic decision for technology transfer to be given serious consideration. There is not enough institutionalized competition to generate alternatives at the middle and final stages of choosing technologies. A number of cases of competitive efforts in development have been mentioned— the design competition for drilling machinery for strip mining, the

VVER versus the RBMK nuclear power reactors, alternative types of compressor equipment for gas pipelines (which incidentally means recourse to different plants and ministries), alternative designs by two different plants for the 500 MW power generating block. Nevertheless, in many of the cases examined there is a strong element of monopoly, in the form of an established set of ideas, and a shortage of institutional and program alternatives. Large trucks are a case in point, and the dependence of the coal industry on machinery plants uninterested in its problems is another. There are four plants producing excavators, but they have deliberately been given a strong intraproduct specialization. The fact that the line of excavators developed at one of the plants was actually assigned by the Ministry to another plant to be produced is an interesting reminder that the customer is really facing one high-level decision maker, not two potential suppliers. This limitation of alternatives is an unsurprising manifestation of the seller's market characteristic of the Soviet-type economies. One of the most interesting things to me is that Minenergo as one big buyer seems to have been more successful than other industries in getting what it wants in the way of technological advances. Minenergo is perhaps the closest analogue in the civilian sector to the Ministry of Defense. For some producers it is essentially the only customer and hence in an advantageous position to exercise control over their innovation efforts and to exert pressure on their performance regarding quality—advantages commonly thought to be important in the Ministry of Defense's generally successful record of technological advance.

At various points in the discussion I have touched on the question of where among the various participants in the R and D process the weakest link is. When we find an example where the technological level seems to be low or technological upgrading proceeds slowly (inadequate equipment for open-pit coal mining, poor reliability or efficiency of gas turbine equipment, technologies poorly oriented to the needs they are supposed to serve), can we point to the performance of any particular class of participants as being the major bottleneck? There is some tendency in the literature on Soviet technology to say that the weakest link in the cycle is at the development end, especially in the behavior of the firms who are supposed to produce equipment embodying new technology.

There is perhaps some support in this study for that idea. Apparently in most of our cases the people responsible for basic research, for making technical and economic evaluations of alternative possibilities, and for designing new technological systems, are able and well trained.

In a number of cases we would have to acknowledge them as bold and inspired. The work on fusion, on MHD, on high-voltage power transmission are cases in point. The testimony of coal industry officials is generally favorable as regards the designers, who have come up with concepts and designs adequate to solve the problems the producers have identified. And certainly in a number of cases we have looked at (especially in the coal industry), it is the ineptitude or disinterest of the equipment producers that has caused the most serious delays and disappointments in upgrading the technical level of the industry. But I think that it is fundamentally untenable to identify this as *the* weak link. In most of the cases presented here, there is a compound failure of responsibility involving many of the actors and a number of the stages of development.

The people who design open-pit mining schemes themselves contributed a prejudice that probably delayed efforts to produce excavator and truck models adapted to the needs of the mines they were operating. The designers of the gas-steam combined cycle prototype made an error in assuming that gas or low-sulfur fuel oil would be available, with the result that the development work done to date on that idea represents, to some extent, time lost on a sidetrack. But I doubt that we can really fault them for such a lack of foresight—they seem to have done their work well within the constraints of the problem as it was assigned to them. Many of the cases suggest that the system pushes the designers and manufacturers into premature efforts on a timetable that does not allow enough preliminary work—as in the decisions to produce 500 MW and 800 MW generating blocks when experience or information necessary for designing boilers that would work on the intended fuels was simply not available. (It is impossible to know to what extent the R and D performers themselves were remiss in not insisting on a timetable that would permit them to do their job right.) Often, the complaint about equipment once it is produced is that it doesn't really fit the conditions encountered, but it sometimes turns out that these conditions were never properly communicated to the designers.

Innovators get caught in a web of interdependencies that frustrate what may well be honest and intelligent efforts to raise the technological level of their outputs or operations. I believe the coal industry planners really have come to accept that trucks and larger excavators could improve the economic conditions of strip mining, but their aspirations in this direction are now constrained by the actions of the truck and excavator producers. The excavator producers in turn say that they cannot obtain steel of the appropriate qualities to meet the specifications the coal industry has articulated. This breakdown through interdependence underlines a thesis that the weaknesses in Soviet innovation probably have less to do with technological knowledge proper than with organization. Of course this is not inconsistent with the way innovation is usually understood—central to Schumpeter's concept is the fact that the innovator has to overcome problems of finance, marketing, labor, and so on, as well as purely technical problems. But one does develop a feeling that these problems are more intractable in the Soviet environment than in a more pluralistic system.

Fundamentally the problem seems to be with the weakness of feedback mechanisms in the Soviet system. We should not expect innovation to work perfectly smoothly in any economy—the penalty of working with the unknown is that some of the expectations that have guided the creators of new technology will be falsified, thus invalidating some earlier decisions and requiring their abandonment or modification. The important thing is that there be mechanisms to generate the kind of adjustments that will adapt original decisions to give second-best solutions. When one arrives at the later stages of the cycle, one is stuck with a framework of settled issues, maybe even design and production commitments, and there is no choice but to keep experimenting until the innovation is made to work one way or another. The peculiar feature of the Soviet system is that this learning often fails to induce the modifications that are so painfully discovered. Dienes says that, despite extended difficulties in making boilers work on Kansk-Achinsk lignites and the resort to extensive modification to make them work, the boiler producers just kept on producing the boilers according to the original designs. Another interesting variation on the same problem has to do with boiler-turbine disproportions. Boilers have been designed to produce steam outputs adequate to match the capacities of turbines on certain expectations about the heat content of fuels. But these expectations about heat content are almost universally unmet, leading to a wasteful underutilization of turbine and generator capacity. Under the existing institutional and incentive system, however, this kind of informational correction is not transmitted back to the attention of the designers. The explanation for all these breakdowns is the well-known failure of a strongly hierarchical structure to provide enough lateral communication or powerful enough leverage to clients to make the interactions between all the participants effective.

Rather than trying to apportion blame among the phases and the participating actors, though, it may be more useful to review our evidence from the point of view of the task of commercialization of technologies and to consider how the Soviet approach to performing this function differs from that in the United States—there are some striking differences. Furthermore, I believe that the different degrees of success the USSR has had in various technical areas are to be sought primarily in the commercialization phase and are related to the differences in the kinds of tasks posed by commercialization in different technical areas.

The distinctive features of Soviet commercialization, as I see them, might be described as follows. A decision to move ahead on introducing a significant innovation is likely to be made only when there is some unusual pressure—an outside example, a crucial bottleneck, a crash program for some sector that creates a specific pressing technological need. In the absence of such pressure, R and D work in any area of technology is likely to mark time, with little effort put into intermediate kinds of applied research, such as accumulation of data or research on materials and components, that would provide a foundation for a subsequent decision to commercialize. Once a commitment is made, however, it is likely to generate a crash program. In this effort stages are telescoped, and attention is given primarily to some components of the system involved, with little attempt to design the whole technology as an optimized system. A fairly ambitious step upward is likely to be attempted in choosing the scale of the demonstration plant, unit size of equipment, or technical parameters. A corollary of this approach is a strong preference for going directly to an expensive prototype or demonstration facility as a substitute for intensive work on the elements of the new technology to enhance predictability in design. Rather, the testing of the new technologies is likely to take place using what are intended essentially as commercial designs, to which the producers have made heavy commitments, or in facilities that are expensive enough that they must be considered for commercial exploitation. Such an overambitious and premature commitment may well result in large wastes and delays.

This description probably sounds as if I had taken the MHD program, about which we happen to know a great deal, and generalized its main features, asserting that it applies to energy R and D in general. But some reflection on the various cases considered shows that it fits quite well the development effort for gas turbine compressors, the creation of each successive generation of power generating equipment, and the experience with excavators and trucks, to mention only the most obvious examples. The report of the U.S. nuclear power reactor delegation that visited the USSR in 1974, underlines the preference for "complex or integrated reactor experiments for test purposes, versus the U.S. preference for single purpose tests," and further elaborates:

Apparently major plant construction was started without complete designs in hand and without extensive prior proof-testing of all features. . . . It is difficult to know if the examples we saw are merely making the best of earlier oversights, but the Soviets clearly accept risks associated with the price of moving forward with demonstration plant construction. [ERDA, 1974, p. 3]

Coal conversion reveals a long history of rather poorly guided effort, followed by the scheduling of an overambitious demonstration effort. The test data and economic evaluations of the processes were based on units handling about 15–20 tons of coal a day. The next two steps, however, are to be a plant of one million tons annual capacity and then one of 24 million tons annual capacity. (As an indication of how large a jump this represents, the largest demonstration plant for coal processing now scheduled in the U.S. approach is a 219 thousand tons per year liquification plant.) Because a step of that magnitude was probably premature, Minenergo has subsequently vacillated in pressing ahead with it, as mentioned in Chapter 6. Slurry pipelines, geothermal and solar power exemplify technologies that have not yet been given the nod, and R and D in those areas pokes along with little indication of exploratory work appropriate to their current status and prospective importance and designed to provide a basis for future decisions about application.

The kind of aimless, misdirected effort that can take place is suggested by the following critique of a coal industry institute's program to utilize coal dust to make smokeless lump fuel. This is presented in the feuilleton style of Soviet investigative journalism and may be a one-sided, but hardly fanciful, account.

Since this use of coal dust was an alluring possibility, the scientists easily obtained money to build a pilot plant.

The most logical location would have been either the Donets Basin or the Kuznetsk Basin. Instead, reasoning that it would be much more pleasant to work amidst the beautiful landscapes of the Northern Caucasus, the scientists decided to build the plant in the village of Kumysh, Karachai-Cherkess Autonomous Province. For developmental purposes they would use coal dust from the mines of Stavropolugol and then later convert to local coals when building full-scale production facilities in the Donets Basin and Kuznetsk Basin.

Time passed, and the newly built plant opened with a flourish of publicity. Since changes had occurred in the institute's staff in the meantime, with some individuals retiring and others taking jobs elsewhere, there were heated discussions over who could perform research. At the Kumysh plant everything remain unchanged: There was smoke, but no smokeless lump fuel.

Representatives of the Karachai-Cherkess People's Control Committee visited the ill-fated plant and saw what the scientists' overeagerness and fantasy-seeking had led to. In their blind haste, they had accepted the production complex before it was finished. The equipment was poorly engineered and inadequate to its intended purpose.

The USSR People's Control Committee investigated next. Again no real progress was made. Instead the institute devised a sham to give the appearance of progress. Abandoning the smokeless lumps, the researchers switched to a new project. They acquired a large shipment of coal from the Kansk-Achinsk Basin and set out to find a way to reduce its characteristic high moisture content. They drove their equipment at full blast until it was worn out. The plant had to completely shut down and major repairs were required.

A technical conference chaired by A. Mazin, Russian Republic Minister of the Fuel Industry, discussed the smokeless lump project and arrived at the conclusion that the project had failed.

It cost the state 1.134 million rubles to build the pilot plant. Five years of operation cost another 911,000 rubles. Now the pilot plant is being converted to a coal-briquetting facility so that this absolutely useless enterprise will at least be kept busy doing something. Naturally the necessary modifications require more investments. [*Current Digest of the Soviet Press*, vol. XXVIII, No. 51, p. 12]

The characterization I have offered may seem to differ in some respects from commonly held ideas, and we should note that it also contains an internal contradiction. A recent study of military R and D by the RAND Corporation concludes that Soviet design tends to be conservative, preferring small incremental changes rather than creation of whole new systems intended to achieve large upward steps in performance. Indeed, our characterization suggests that the ambitiousness of the step upward may be offset by conscious avoidance of developing the new generation as a complete system. To insure against failure, the energy equipment programs we have looked at tend to use as auxiliary equipment models that have already been extensively tested in practice. The innovative elements are usually introduced into a fairly conventional environment, and in this way the R and D risk can be limited to the novel elements. Recall that in the U-25 MHD plant, everything outside the MHD element itself is quite conventional. The aviation-type compressors for gas line use were based on an existing aviation engine, and the task of designing a new model from scratch was rejected. The heavy emphasis on parts commonality evidences the same philosophy in the sphere of individual items of equipment. The 300 MW power block was reported to have 65 percent parts commonality with early models, and we find the same emphasis on parts commonality in the excavator field and in the development of underground mining machinery. This form of incrementalism is a way of easing the commercialization of a new technology at the crucial sticking point—the manufacture of new equipment—by using components already in production.

As another form of insurance, early versions and demonstration facilities are commonly designed with considerable reserves to cope with disappointments and to permit subsequent increases in capacity once operating experience is acquired. The first 500 KV line was operated originally at 400 KV; only later were experiments made to see if it could operate at 500 KV. As another example it is reported that the original design for the 1,000 MW RBMK nuclear reactor was so conservative that the upgrading to a 1,500 MW version can be achieved just by intensifying the heat exchange process, without modifying the design of the reactor as a whole. Likewise, in the VVER nuclear reactor, capacity is to be raised by exploiting the technical slack in the original model revealed through operating experience, rather than by extensive redesign of the reactor (*Atomnaia nauka i tekhnika v SSSR*, pp. 37, 43). But even with this kind of insurance, Soviet experience with new power generating blocks (especially in boiler design), the MHD facility, the nuclear power program, and several of the excavator models, all suggest that, in relation to the actual capabilities of the R and D establishment, planners decided on big enough steps into the unknown to cause real difficulties.

I believe that the R and D approach outlined above has some serious disadvantages. U.S. observers of the MHD program seem to agree that the Soviet method is wasteful, since it fixes too many parameters in advance, puts ideas into metal and concrete too soon. And, as mentioned in Chapter 7 in connection with development of turbine compressors, Soviet commentators themselves are often highly critical of the failure to do enough testing as new equipment is being designed and readied for production. Certain aspects of the approach, however, may have important offsetting advantages, and the point here is not that Soviet R and D practices are inept, but that they are different from ours. It seems to me fairly obvious that the most helpful explanation for the differences from U. S. practices is the absence of the competitive pressures that influence R and D efforts in the market economy, especially in the private sector. In the United States, one of the main reasons for so much caution in the design and testing phase is the desire of the developer of new equipment that it be competitive and appealing to prospective customers when it is ready to be marketed. Potential users will want to know when it can be delivered; the seller of the equipment must be able to assure buyers about cost and performance. Failure here can mean that a firm's innovation will not win acceptance. The same argument applies to a considerable extent even to the energy R and D programs sponsored or financed by governments rather than profit-making firms. In most market economies any new technologies developed with the stimulus of government financing will still have to meet market tests if they are to be successfully commercialized. Moreover, a technology brought successfully to the commercial stage in a given national context can be wiped out by international competition even if it is technically successful. As spectacular examples, consider the Concorde and the Dragon reactor.

In the Soviet setting many of those considerations do not matter. The user ministries have much more limited motivation and freedom to reject technologies that are not commercially appealing than do firms in a market economy. If the decision from the top is that the new technology is to be developed in variant A rather than variant B then there is enough monopoly in the system to foreclose the subsequent emergence of a truly superior form B to undercut the effort invested in A. Considerations of international competitiveness also appear much less forcefully in the USSR, and in a different way. Foreign competition is unlikely to supersede a domestically developed technology, except possibly in the form of technology transfer if the domestic program is a flagrant failure. International competitiveness may also influence the fate of the new technology if its development is motivated in part by the hope for foreign sales of the equipment (as in the example of power-generating equipment). This kind of international sensitivity was generally unimportant in the USSR in the past, though it has become more prevalent in recent years. There is now far more frequent resort to technology transfer; the USSR now has had some success and has larger ambitions regarding exports of energy technology. Eastern Europe, which in the past was essentially a captive market for the USSR, is now a much more discriminating comparison shopper in choosing the source of new energy technology. Students of the Soviet-type economies have often remarked on the potential virtues of opening up the Soviet economy as a way of disciplining domestic producers, and we have here an interesting variation on that idea—i.e., that more involvement with the world economy has the potential to transform somewhat even the R and D processes that precede production.

Another way to interpret this difference in Soviet and Western approaches to R and D is to relate it to differences in the kinds of uncertainties that dominate the respective systems. It was suggested earlier that many of the failures and delays noted in the various programs flow from the network of interdependencies, in which the various actors find themselves with very limited power to command information or to influence behavior of those on whom they depend in carrying out their part of the effort. In the United States, R and D program managers work on reducing technological uncertainties by precommercialization forms of R and D, the better to be prepared to deal with competitive uncertainties involved in winning acceptance at the commercial stage. In the USSR, the uncertainties imposed by interorganizational unpredictability are the ones that can doom a project, and program managers try to ease these by demanding only conventional inputs from outside the program. At the same time they are much less worried about getting the technology accepted once they have developed it, and so they push ahead with a crash program to deliver it in some kind of form, not worrying that they will have to do considerable backtracking afterward to cope with defects that emerge in practice and that were not thought about enough in the precipitate rush to commercialization.

The prospects for significant improvements in the interorganizational obstacles to improving Soviet R and D through more lateral interaction and responsiveness are probably not great, and the likelihood that the distinctive R and D patterns we have seen will retain their vitality is high. Many experiments have been tried to make R and D organizations more responsive to the needs of clients and to regroup organizations into new aggregations that would break down internal barriers. The shift to unified fund financing and to the creation of scientific-production associations (NPO) are examples of these efforts described in Chapter 2. But we should not be misled into believing that behavior will change significantly or that on balance the improvements from new structures will much exceed the defects that they bring with them. An official of the Soviet Committee for Science and Technology (GKNT) well acquainted with R and D in the electric power sector asserts that after five years under the new unified-fund system, though it has been incessantly lauded in the general literature as a gigantic step forward, there was no substantial change in the way R and D was performed and that the system should be abandoned (L. A. Vaag in *Voprosy ekonomiki*, 1975:8, pp. 121–122). In an effort to break down departmental barriers in the R and D process and increase sensitivity to client needs in the oil field equipment field, one of the scientific production associations called Soiuzneftemash has been created. Combining R and D organizations, experimental production and test facilities, and production plants, this organization also includes units whose responsibility is to follow up machinery deliveries with field service and testing. It sounds just like what the critics of Soviet R and D have recommended, but it is interesting to find the leaders of an R and D organization within Soiuzneftemash complaining that in fact the new arrangement has blocked effective work on new equipment:

> The Association has begun to assign small urgent tasks to the Institute which have no direct relation to either the Institute's planned program or specialization. They are mainly technical problems arising at different plants of the Association, which should be dealt with by plant technical personnel. Since January 1976, the Institute has received almost 600 such requests. It has been almost deprived of its experimental base, which is being largely used for mass production. Its research is seriouly hampered, and it is being turned into a maintenance brigade [*Sotsialisticheskaia Industriia*, 2 February 1977, p. 2]

One of the interesting questions is how to explain variations in R and D effectiveness within the energy sector, such as the difference between the electric power field (where the R and D process for all its faults has solved many significant problems) and the coal industry (where lavish R and D resources have apparently been powerless to raise the technological level anywhere near that in the capitalist countries). Several differences are probably important in the explanation. Relative resource scarcities may have played a part. In the coal industry much innovation involves capital-for-labor substitution, and an abundant labor supply simply meant little pressure to innovate in this way. In the electric power sector, a need to save capital and fuel may have spurred innovation. As suggested in Chapter 3, fuel is relatively expensive in the USSR, and since electric power generation is the major fuel consuming sector, it has naturally stimulated serious R and D effort aimed at cutting the heat rate. Conditions in the electric power industry also offer large scope for following a few simple strategic principles to raise productivity, an approach toward which Soviet planning is very partial, and which is easily seen in the programs we have de-scribed for this industry. It is more nearly possible in electric power than in coal to design a few basic pieces of equipment, such as a 300 MW power block to be installed in a standard building, and then to replicate this design all over the USSR. In the coal industry separate solutions are needed for thick and thin seams, for flat-lying and sloping seams, for underground and strip mining, and so on. But part of the explanation must lie in the kind of market-power considerations dis-cussed earlier. Minenergo has a more powerful position in relation to its technology suppliers than do many other organizations in the energy sector. Most of its suppliers work for Minenergo almost exclusively, whereas many of the plants supplying equipment to producers in the other energy branches serve a bigger variety of customers. I also sus-pect that Minenergo has an unusual situation in its large project-mak-ing organizations—it has a very powerful say in the design of nuclear power stations, for example, and hence a strong influence on nuclear R and D organizations in machinery industries that supply the ma-chinery that determines their technological level.

It might be thought that an in-depth look at Soviet experience in dealing with energy R and D might generate some predictions as to how Soviet energy policy will evolve, how technological considerations will influence outcomes on the big choices confronting Soviet energy policymakers mentioned throughout this book.* In fact, such a hope was a part of the inspiration for this study. As my investigation has proceeded, however, that expectation has become illusory, and I would like to conclude with some summary reflections on the general issue of how energy R and D interacts with energy policy.

It seems unlikely that the USSR will achieve the kind of dramatic breakthroughs in any given technological area that could tilt the bal-ance of economic advantage decisively one way or the other on any of the major choices. Most Soviet energy R and D programs drag out over long periods, the equipment and facilities that finally emerge from them only partially meet the needs that motivated them. This is inherent in R and D, of course, and might be said about any econ-omy, but I read the Soviet record as more disappointing in this respect than the norm in market economies. In the light of past experience, it just doesn't seem realistic to expect the kind of great success in devel-oping the technology of coal conversion or in power transmission that

*For a more extensive description of the choices inherent in the Soviet energy situation, see the excellent treatment in Shabad and Dienes (1979).

would give Kansk-Achinsk coal a dominant place in the solution of the energy deficit in the European USSR. Nor can one expect that the nuclear program, even in the light of its efforts to adapt reactors to permit applications other than power generation, will proceed, in accordance with the hopes of its managers, to make a dominating contribution to solving that problem.

By a similar argument, we should not expect that Soviet R and D in the energy sector is likely to so influence the structure of relative advantage in energy options as to move the main features of Soviet policy much away from that in the rest of the world. There are and will continue to be many distinctive features in the pattern of production and utilization of energy resources in the USSR, but these grow much more out of the nature of the Soviet resource endowment and demand structure than out of special Soviet strengths or predilections in the creation of energy technologies. It is true that Soviet planners have made some major commitments in energy technology that differ from those in the United States, such as the emphasis on the breeder reactor or the dominant place given the turbodrill. But it is still too early for a conclusion that they will succeed in making the breeder (especially when we remember the problems it entails in other parts of the fuel cycle) an environmentally viable and economic contributor to overall energy supply. And the turbodrill illustrates the phenomenon, fairly common in Soviet energy R and D experience, of pulling back from a distinctive technological line if it runs counter to world technological trends.

From the other side, none of the technological obstacles associated with the various energy policy alternatives are likely to be so insurmountable as to render a given option impossible. If domestic R and D efforts are unavailing in finding an acceptable solution to some crucial aspect of a given energy option (say in raising the productivity of gas pipelines), technology transfer often offers a way out. Our study suggests that that choice will be made only under the pressure of extreme urgency, but it offers a flexibility that heavily undercuts the notion that differential success among technological areas will be determinative in resolving the big issues of energy policy.

One of the distinctive elements differentiating the world view of economists from that of other students of social processes is a sensitivity to the fact that there are always more alternatives than appear to be available at first blush, that there are all kinds of substitutions and alternatives that make it very short-sighted to draw conclusions about what can and cannot be done. What is fascinating about R and D, what makes it so elusive to analyze from the point of view of making economic choices, is that it expands the range of choice still further. Technological alternatives aren't given, they are created, and the way they are created is through R and D efforts. It does not make much sense to say that the breeder reactor won't work, without specifying the particular technological embodiment of the general physical principle involved. Because R and D in its broadest concept involves the possibility of going back and starting with more fundamental approaches to almost *any* problem, the range of alternatives at any one place along the continuum of technological possibilities can be increased by shifting the effort backward toward the basic research end of the spectrum.

The following generalization about Soviet energy R and D may be helpful as we try to predict what may happen as Soviet energy policy unfolds. What is distinctive about the Soviet system, what constitutes the real weakness in the R and D process (which is one of the means to be manipulated in any large policy decision area) is its general clumsiness in its movement back and forth along the R and D spectrum—it makes commitments too soon, generates too few technological alternatives at each stage, finds it difficult to move backward to correct the consequences of miscalculations once commitments are made. This tendency is deeply ingrained in the system, perhaps an inherent feature of its basic organizational properties, and has a pervasive inference on technological advance. An awareness of that tendency is perhaps the best general orientation this study can provide for our continuing effort to evaluate technological potential as we seek to forecast Soviet energy policy.

Bibliography

BOOKS & ARTICLES

Abramov, B. S. 1959. *Energetika semiletki* [Electric Power in the Seven Year Plan]. Moscow.

AEC (U.S. Atomic Energy Commission). 1970. Division of Reactor Development and Technology, *Soviet Power Reactors—1970*. Washington.

Akademiia Nauk SSSR, Spravochnik [Academy of Sciences of the USSR, Handbook]. Moscow, 1975.

Alekseev, G. N. 1978. *Prognoznoe orientirovanie razvitiia energoustanovok* [Predicted Direction of Development of Electric Power Installations]. Moscow.

Altshuler, V. S. 1976. *Novye protsessy gazifikatsii tverdogo topliva* [New Processes for Gasifying Solid Fuel]. Moscow.

Amann, R., et al. 1977. *The Technological Level of Soviet Industry*. New Haven: Yale University Press.

Amiian, V. A., et al. 1962. *Tekhnicheskii progress v dobyche nefti* [Technical Progress in Oil Production]. Moscow.

ANSSSR (Academy of Sciences of the USSR). 1968. *Razvitie otkrytykh gornykh rabot v SSSR* [Development of Open-Pit Mining Operations in the USSR]. Moscow.

ANSSSR. 1973. *Rukovodiashchie ukazaniia k ispol'zovaniiu zamykaiushchikh zatrat na toplivo i elektricheskuiu energiiu* [Guidelines for Using Marginal Costs for Fuel and Electric Power]. Moscow.

ANSSSR, Institut ekonomiki, 1969. *Nauchno-tekhnicheskii progress i khoziaistvennaia reforma* [Scientific-Technical Progress and Economic Reform]. Moscow.

Arkhangel'skii, V. N. 1976. *Planirovanie i finansirovanie nauchnykh issledovanii* [Planning and Financing of Scientific Research]. Moscow.

Atomnaia nauka i tekhnika v SSSR [Atomic Science and Technology in the USSR]. Moscow, 1977.

Atomnoi energetiki XX let [Nuclear Power is Twenty Years Old]. Moscow. 1974.

Avrukh, A. Ia. 1966. *Problemy sebestoimosti elektricheskoi i teplovoi energii* [Problems of the Cost of Electrical and Thermal Energy]. Moscow.

Avrukh, A. Ia. 1977. *Problemy sebestoimosti i tsenoobrazovaniia v elektroenergetike* [Problems of Cost and Price Formation in Electric Power]. Moscow.

Baibakov, N. K., ed. 1972. *Gosudarstvennyi piatiletnyi plan razvitiia narodnogo khoziaistva SSSR na 1971–1975 gody* [The State Five Year Plan for Development of the National Economy of the USSR for 1971–1975]. Moscow.

Batov, V. V., and Iu. I Koriakin. 1969. *Ekonomika iadernoi energetiki* [Economics of Nuclear Power]. Moscow.

Bazarova, G. V. 1974. *Pribyl' i khozraschet v usloviiakh nauchno-tekhnicheskogo progressa* [Profit and Economic Accountability in Conditions of Scientific-Technical Progress]. Moscow.

Belan, R. 1962. *Perspektivy razvitiia chernoi metallurgii SSSR* [Prospects for the Development of Ferrous Metallurgy in the USSR]. Moscow.

Bergson, Abram, and S. Kuznets. 1963. *Economic Trends in the Soviet Union.* Cambridge, Mass.: Harvard University Press.

Bergson, Abram. 1964. *The Economics of Soviet Planning.* New Haven: Yale University Press.

Beschinskii, A. A., and Iu. M. Kogan. 1976. *Ekonomicheskie problemy elektrifikatsii* [Economic Problems of Electrification]. Moscow.

Bituminous Coal Facts. 1972.

Bliakhman, L. S. 1968. *Voprosy ekonomiki i planirovaniia nauchnykh issledovanii* [Questions of Economics and Planning of Scientific Research]. Leningrad.

Bobolovich, V. N. 1974. "Rol' bystrykh reaktorov v strukture razvyvaiushcheisia sistemy iadernoi energetiki" [The Role of Fast-Neutron Reactors in the Structure of the Developing Nuclear Power System]. *Atomnaia energiia,* 1974:4.

Bocek, K. 1974. "Our Uranium and the USSR." *Listy.* Reported in *Radio Liberty Dispatch,* 29 July 1974.

Bogopol'skaia, T. I. 1979. *Tekhniko-ekonomicheskie pokazateli gazotransportnoi sistemy Zapadnoi Sibiri* [Technical-Economic Indicators of the Gas Transport System of Western Siberia]. Moscow.

Bol'shaia Sovetskaia Entsiklopediia [The Great Soviet Encyclopedia], 2nd Ed.

Boncher, W. 1976. *Innovation and Technical Adaptation in the Russian Economy: The Growth in Unit Power of the Russian Mainline Freight Locomotive.* Ph.D. Dissertation, Indiana University.

Bratchenko, B. F. 1960. *Perspektivy razvitiia ugol'noi promyshlennosti SSSR* [Prospects for the Development of the Coal Industry of the USSR]. Moscow.

Campbell, R. W. 1968. *The Economics of Soviet Oil and Gas.* Baltimore: The Johns Hopkins University Press.

Campbell, R. W. 1974. *The Soviet-Type Economies.* Boston: Houghton Mifflin.

Campbell, R. W. 1976. *Trends in the Soviet Oil and Gas Industry.* Baltimore: The Johns Hopkins University Press.

Campbell, R. W. 1978a. *Soviet Energy R and D: Goals, Planning, and Organizations.* The RAND Corporation, R-2253-DOE.

Campbell, R. W. 1978b. *Soviet Energy Balances.* The RAND Corporation, R-2257-DOE.

Campbell, R. W. 1979. *Basic Data on Soviet Energy Branches.* The RAND Corporation, N-1332-DOE.

Chernukhin, A. A., and Iu. N. Flakserman. 1975. *Ekonomika energetiki SSSR* [Economics of Electric Power in the USSR]. Moscow.

Chikradze, Sh. G. 1972. *Geotermiia Kolkhidskoi nizmennosti* [Geothermal Conditions in the Kolkhid Depression]. Moscow.

Christenson, C. 1962. *Economic Redevelopment in Bituminous Coal.* Cambridge, Mass.: Harvard University Press.

CIA (U.S. Central Intelligence Agency). 1965. *Comparison of Power Plant Technology and Costs in the USSR and the U.S.* Washington, 1965.

CIA. 1975. *Soviet Long-Range Energy Forecasts.* Washington.

CIA. 1976. *Ruble/Dollar Ratios for Construction.* Washington.

CIA. 1977. *Handbook of Economic Statistics.* Washington.

Darmstadter, J. 1971. *Energy in the World Economy.* Baltimore: The Johns Hopkins University Press.

Dobva, A. S. 1973. *Tekhnicheskii progress i trud na predpriiatiiakh po dobyche uglia* [Technical Progress and Labor in Coal-mining Enterprises]. Moscow.

Dombrovskii, N. G. 1969. *Ekskavatory* [Excavators]. Moscow.

Dronov, F. A., and M. N. Shatokhinaia. 1970. *Ekonomika osvoeniia novoi produktsii* [Economics of Mastering New Products]. Minsk.

Dubinskii, V. G. *Ekonomika razvitiia i razmeshcheniia nefteprovodnogo transporta v SSSR* [Economics of Development and Location of Oil Pipeline Transport in the USSR]. Moscow.

Dvorov, I. M. 1976. *Geotermal'naia energetika* [Geothermal Power]. Moscow.

ECE (Economic Commission for Europe). *Annual Bulletin of Coal Statistics for Europe.*

Edison Electric Institute. 1970. *Historical Statistics of the Electric Utility Industry.* New York.

Edison Electric Institute. *Statistical Yearbook.*

EEC (European Economic Community). *Energy Statistics Yearbook.*

Emelianov, I. Ia. 1975. *Upravlenie i bezopasnost' iadernykh energeticheskikh reaktorov* [Control and Safety of Nuclear Power Reactors]. Moscow.

Energeticheskaia, atomnaia, transportnaia i aviatsionnaia tekhnika. Kosmonavtika [Electric Power, Nuclear, Transport and Aviation Technology. Cosmonautics]. Moscow, 1969.

Energeticheskoe mashinostroenie [Electric Power Machine Building, 1917–1967]. Moscow, 1967.

Energomashinostroenie Leningrada v 1959–1965 gg [Electric Power Machine Building in Leningrad in 1959–1965]. Leningrad, 1958.

ENIN (Krzhizhanovskii Power Institute). 1968. *Vyravnivanie grafikov nagruzki energeticheskikh sistem i vybor tipa elektrostantsii dlia pokrytiia pikovykh nagruzok* [Evening out the Load Curves of Power Systems and Choice of Power Station Type for Covering Peak Loads]. Moscow.

ENIN. 1963. *Metody pokrytiia pikov elektricheskoi nagruzki* [Methods of Covering Electric Load Peaks]. Moscow.

ERDA (U.S. Energy Research and Development Administration): Division of Reactor Research and Development. 1974. *Soviet Power Reactors—1974.* Washington.

ERDA. 1976. *A National Plan for Energy Research, Development and Demonstration: Creating Energy Choices for the Future.* Volume I: The Plan. Washington.

FEA (U.S. Federal Energy Administration). 1977. *Energy and U.S. Agriculture, 1974 Data Base:* two volumes. Washington.

FPC (U.S. Federal Power Commission). 1973. *Natural Gas Survey.* Washington.

FPC. *Steam-Electric Plant Construction Cost and Annual Production Expenses.*

FPC. *Gas Turbine Electric Plant Construction Costs and Annual Production Expenses.*

FPC. *Electric Power Statistics.*

FPC. *Statistics of Privately-owned Electric Utilities.*

FPC. 1977. *National Gas Flow Patterns, 1975.* Washington.

FPC. *Statistics of Interstate Natural Gas Pipeline Companies.*

Friman, R. E. 1976. *Magistral'nye truboprovody* [Transmission Pipelines]. Moscow.

Furman, I. Ia. 1978. *Ekonomika magistral'nogo transporta gaza* [Economics of Pipeline Transmission of Gas]. Moscow.

Galkin, L. G. 1966. *Rezervy proizvodstva v dobyche nefti i gaza* [Production Reserves in Extraction of Oil and Gas]. Moscow.

Gal'perin, V. M. 1975. *Razvitie i perspektivy transporta gaza* [Development and Prospects for Gas Transport]. Moscow.

Gerasimov, B. Ia. 1969. *Perekachivaiushchie agregaty dlia magistral'nykh gazoprovodov* [Compression Equipment for Gas Transmission Lines]. Leningrad.

Gidroproekt. 1972. *Tezisy dokladov i soobshchenii na vtoroi nauchno-tekhnicheskoi konferentsii gidroproekta* [Theses of Reports and Communications at the Second Scientific-Technical Conference of Gidroproekt]. Moscow.

Gitelman, L. D. 1974. "Obosnovanie normativnogo sroka sluzhby energoblokov AES." [Justification of the Normal Service Life of Nuclear Power Units]. *Atomnaia energiia*, 1974:4.

Gorshkov, A. E., ed. 1967. *Ekonomika topliva na elektrostantsiiakh i v energosistemakh* [Economy of Fuel in Power Stations and Systems]. Moscow.

Gosplan SSSR. 1974. *Metodicheskie ukazaniia k razrabotke gosudarstvennykh planov razvitiia narodnogo khoziaistva SSSR* [Methodological Guidelines for Developing State Plans for Development of the National Economy of the USSR]. Moscow.

Gregg, D. W. 1976. *An Overview of the Soviet Effort in Underground Gasification of Coal.* Lawrence Livermore Laboratory, UCRL-52004.

Gregory, Paul, and Robert Stuart. 1974. *Soviet Economic Structure and Performance.* New York: Harper and Row.

Grishaev, E. 1970. In SEV (Council for Mutual Economic Assistance). *Upravlenie, planirovanie i organizatsiia nauchnykh i tekhnicheskikh issledovanii* [Management, Planning and Organization of Scientific and Technical Studies]. Moscow.

Gvishiani, D. M. 1973. *Osnovnye printsipy i obshchie problemy upravleniia naukoi* [Basic Principles and General Problems of Managing Science]. Moscow.

Iatskov, V. S. 1976. *Ekonomicheskie problemy tekhnicheskogo progressa v ugol'noi promyshlennosti UkrSSR* [Economic Problems of Technical Progress in the Coal Industry of the Ukrainian SSR]. Kiev.

Ivasenko, V. N. 1977. *Tsena i kachestvo energeticheskikh uglei* [Price and Quality of Steam Coals]. Moscow.

Janes' All the World's Aircraft. New York: Franklin Watts.

Kaikkonen, Kh. and P. Silvennoinen. 1976. "Plutonievaia zagruzka VVER-440" [Loading the VVER-440 with Plutonium]. *Atomnaia energiia* [Nuclear Energy], vol. 40, 1976:4, pp. 283–286.

Karnaev, V. D. 1972. *Sovremennaia nauchno-tekhnicheskaia revoliutsiia* [The Contemporary Scientific-Technical Revolution]. Moscow.

Karol', L. A. 1975. *Gidravlicheskoe akkumulirovanie energii* [Hydraulic Storage of Power]. Moscow.

Khaskin, G. Z. 1975. *Osnovnye fondy gazovoi promyshlennosti* [Fixed Assets of the Gas Industry]. Moscow.

Kirillin, V. A. 1974a. *O perspektivakh MGD elektrostantsii v energetike* [On the Prospects for MHD Stations in Electric Power]. Moscow.

Kirillin, V. A. 1974b. *Energetika budushchego* [Electric Power of the Future]. Moscow.

Kirillin, V. A. 1976. "Design Considerations for the First Commercial Scale MHD Generation Unit." In *Third U.S.-USSR Colloquium on Magneto-hydrodynamic Electrical Power Generation*. ERDA, CONF-761015.

Kortunov, A. K. 1967. *Gazovaia promyshlennost' SSSR* [The Gas Industry of the USSR]. Moscow.

Koshkarev, A. P. 1972. *Problemy ekonomicheskoi podgotovki proizvodstva* [Problems of Economic Preparation of Production]. Kiev.

Kosov, E. V., and G. Kh. Popov. 1972. *Upravlenie mezhotraslevymi nauchno-tekhnicheskimi programmami* [Management of Interbranch Scientific-Technical Programs]. Moscow.

Krapchin, I. P. 1976. *Effektivnost' ispol'zovaniia uglei* [Effectiveness of Utilization of Coal]. Moscow.

Kruger, Paul, and Carol Otte. 1973. *Geothermal Energy*, Palo Alto: Stanford University Press.

Kulikov, A. G. 1964. *Ekonomicheskie problemy uskoreniia tekhnicheskogo progressa v promyshlennosti* [Economic Problems of Accelerating Technical Progress in Industry]. Moscow.

Kuznetsov, K. K. 1971. *Ugol'nye mestorozhdeniia dlia razrabotki otkrytym sposobom* [Coal Fields for Development by Stripping]. Moscow.

Leningradskaia promyshlennost' za 50 let [The Industry of Leningrad during Fifty Years]. Leningrad, 1967.

Leonkov, A. M. 1974. *Spravochnoe posobie: teploenergetika elektricheskikh stantsii* [Handbook on Thermal Power of Power Stations]. Minsk.

Levental', A. B., and L. A. Melent'ev. 1961. *Tekhniko-ekonomicheskie osnovy razvitiia teplofikatsii v energosistemakh* [Technical-Economic Bases of Development of Cogeneration in Power Systems]. Moscow-Leningrad.

Loginov, V. P. 1971. *Ekonomicheskie problemy tekhnicheskogo progressa v dobyche mineral'nogo syr'ia* [Economic Problems of Technical Progress in Mining Minerals]. Moscow.

Makarov, A. A., and L. A. Melent'ev. 1973. *Metody issledovaniia i optimizatsii energeticheskogo khoziaistva* [Methods of Studying and Optimizing the Electric Power Sector]. Novosibirsk.

Mel'nikov, N. N. 1972. "The Soviet Union—Recent and Future Developments in Surface Coal Mining." *The Canadian Mining and Metallurgical Bulletin*, October, 1972.

Mel'nikov, N. V. 1957. *Razvitie otkrytoi ugledobychi v SSSR* [Development of Open-Pit Coal Mining in the USSR]. Moscow.

Mel'nikov, N. V. 1966. *Mineral'noe toplivo* [Mineral Fuel]. Moscow.

Mel'nikov, N. V. 1968a. *Sovremennoe sostoianie gornoi nauki v SSSR* [The Contemporary Condition of Mining Science in the USSR]. Moscow.

Mel'nikov, N. V., ed. 1968b. *Toplivno-energeticheskie resursy* [Fuel and Power Resources]. Volume I of I. T. Novikov, ed. *Energeticheskie resursy SSSR*: two volumes. Moscow.

Meshkov, A. G. 1976. "Prospects of Development of Fast Nuclear Power Reactors in the USSR." In European Nuclear Conference, Paris, 1975. *Nuclear Energy Material*, Vol. II, Pergamon Press.

Meyerhoff, A. A. "Soviet Petroleum—History, Technology, Geology, Reserves, Potential, and Policy." Forthcoming in AAG Project on Soviet Natural Resources in the World Economy.

Ministerstvo Geologii SSSR (Ministry of Geology of the USSR). 1968. *50 let Sovetskoi geologii* [Fifty Years of Soviet Geology]. Moscow.

Ministerstvo ugol'noi promyshlennosti, NIIPKI po dobyche poleznykh isko-paemykh otkrytym sposobom. 1969. *Bestransportyne sistemy razrabotki mestorozhdenii* [Transportless Systems for Working Deposits]. Cheliabinsk.

Murav'ev, V. M. 1971. *Sputnik neftianika* [The Oil Worker's Companion]. Moscow.

Muravlenko, V. I., ed. 1977. *Sibirskaia neft'* [Siberian Oil]. Moscow.

National Science Board. 1975. *Science Indicators.* Washington.

Nekrasov, A. M., and M. G. Pervukhin. 1977. *Energetika SSSR v 1976–1980 godakh* [Soviet Electric Power in 1976–1980]. Moscow.

Neporozhnyi, P. S. 1970. *Elektrifikatsiia SSSR* [Electrification of the USSR]. Moscow.

Neumann, Jan. 1972. "Vliianie atomnoi energetiki na strukturu toplivno-energeticheskogo balansa ChSSR" [The Influence of Nuclear Power on the Structure of the Fuel and Energy Balance of the Czechoslovak Soviet Socialist Republic]. *Ekonomicheskoe sotrudnichestvo stran-chlenov SEV* [Economic Cooperation of the Member Countries of the Council for Mutual Economic Assistance], 1972:2, pp. 17–21.

Nimitz, N. 1974. *The Structure of Soviet Outlays on R and D in 1960 and 1968.* The RAND Corporation, R-1207-ODRE.

Nolting, L. 1973. *Sources of Financing the Stages of the Research, Development, and Innovation Cycle in the USSR.* U.S. Department of Commerce, Foreign Demographic Analysis Division, Foreign Economic Report No. 3. Washington.

Nolting, L. 1976a. *The Financing of Research, Development, and Innovation in the USSR by Type of Performer.* U.S. Department of Commerce, Foreign Economic Report No. 9. Washington.

Nolting, L. 1976b. *The 1968 Reform of Scientific Research, Development, and Innovation in the USSR.* U.S. Department of Commerce, Foreign Economic Report No. 11. Washington.

Novikov, I. T. 1962. *Razvitie energetiki i sozdanie edinoi energeticheskoi sistemy SSSR* [Development of Electric Power and Creation of the Unified Power System of the USSR]. Moscow.

Novikov, I. T. 1968. *Energeticheskie resursy SSSR* [Power Resources of the USSR]. Moscow.

NSF (National Science Foundation). *Reviews of Data on Science Resources, No. 26. Energy and Energy-Related R and D Activities of Federal Installations and Federally Funded Research and Development Centers.* NSF 76-304.

NSF. *Research and Development in Industry, 1974.* NSF 76-322.

NSF. *Analysis of Federal R and D Funding by Function, Fiscal Year 1969–1977,* NSF 76-325, September, 1976.

Orudzhev, S. A. 1976. *Gazovaia promyshlennost' po puti progressa* [The Gas Industry on the Path of Progress]. Moscow.

Osnovnye napravleniia razvitiia narodnogo khoziaistva SSSR na 1976–1980 gody [Guidelines for the Development of the National Economy of the USSR in 1976–1980]. Moscow, 1975.

Oznobin, N. M., and A. S. Pavlov. 1975. *Kompleksnoe planirovanie nauchno-tekhnicheskogo progressa* [Complex Planning of Scientific-Technical Progress]. Moscow.

Pavlenko, A. S., and A. M. Nekrasov. 1972. *Energetika SSSR v 1971–1975 godakh* [Soviet Electric Power in 1971–1975]. Moscow.

Petrosiants, A. M. 1976. *Sovremennye problemy atomnoi tekhniki v SSSR* [Contemporary Problems of Nuclear Technology in the USSR], 3rd ed. Moscow.

Polach, J. 1968. "Nuclear Industry in Czechoslovakia: A Study in Frustration." *Orbis*, Fall, 1968.

Poliak, A. M. 1965. *Razvitie sortamenta chernykh metallov SSSR* [Development of the Mix of Ferrous Metals in the USSR]. Moscow.

Popov, V. M., et al. 1974. *Energeticheskoe ispol'zovanie frezernogo torfa* [Use of Milled Peat for Power Generation]. Moscow.

Pruzner, S. L. 1969. *Teploenergetika SSSR* [Thermal Electric Power in the USSR]. Moscow.

Ravich, M. A. 1974. *Gaz i ego primenenie v narodnom khoziaistve* [Gas and its Utilization in the National Economy]. Moscow.

Ravich, M. V. 1977. *Effektivnost' ispol'zovaniia topliva* [Effectiveness of Fuel Use]. Moscow.

Resheniia Partii i pravitel'stva po khoziaistvennym voprosam [Decisions of the Party and the Government on Economic Questions]. Moscow, various years.

Rippon, Simon. 1975. "Fast Reactor Progress in the Soviet Union." *New Scientist*, 4 December 1975.

Rosenfel'd, S. Ia. 1961. *Istoriia mashinostroeniia SSSR* [History of Machine-building in the USSR]. Moscow.

Rubinov, N. Z. 1977. *Ekonomika truboprovodnogo transporta nefti i gaza* [Economics of Pipeline Transport of Oil and Gas]. Moscow.

Rudins, G. 1974. *U.S. and Soviet MHD Technology: A Comparative Overview*. The RAND Corporation, R-1040-ARPA.

Semenova, B. N. 1977. *Ekonomika stroitel'stva magistral'nykh truboprovodov* [Economics of Constructing Major Pipelines]. Moscow.

Shabad, T. 1970. *Basic Industrial Resources of the USSR*. New York: Columbia University Press.

Shcherban', A. N. 1969. *Istoriia tekhnicheskogo razvitiia ugol'noi promyshlennosti Donbassa* [History of the Technical Development of the Coal Industry of the Donbass]: two volumes. Kiev.

Shvets, I. T. 1970. *Elektroenergetika piatiletki* [Electric Power in the Five Year Plan]. Kiev.

Sivakov, E. R. 1968. *Tekhniko-ekonomicheskie problemy elektrogeneriruiushchego oborudovaniia energosistem* [Technical-Economic Problems of Power Generation Equipment for Power Systems]. Leningrad.

Smekhov, V. K. 1970. *Ugledobyvaiushchii kompleks KM-87* [The KM-87 Coal-mining Complex]. Moscow.

Smirnov, V. A. 1975. "Gazovaia promyshlennost'." *EKO*, 1975:5.

Smith, Hedrick, 1976. *The Russians*. New York: Ballantine Books.

Smoldyrev, A. E. 1967. *Gidravlicheskii i pnevmaticheskii transport v metallurgii i gornom dele* [Hydraulic and Pneumatic Transport in Metallurgy in Mining]. Moscow.

Smoldyrev, A. E. 1970. *Truboprovodnyi transport* [Pipeline Transport]. Moscow.

Sominskii, V., and L. Bliakhman. 1972. *Ekonomicheskie problemy povyshe-*

niia effektivnosti nauchnykh razrabotok [Economic Problems of Raising the Effectiveness of Scientific Development]. Leningrad.

Spivakovskii, L. I. 1967. *Ekonomika trubnoi promyshlennosti SSSR* [Economics of the Pipe Industry of the USSR]. Moscow.

Spivakovskii, L. I. 1975. *Ekonomika trubnoi promyshlennosti SSSR* [Economics of the Pipe Industry of the USSR]. Moscow. There are two editions: 1967 and 1975.

Sporn, Philip. 1968. "USSR lags far behind in power race with U.S." *Electrical World*, November.

Statistical Abstract of the U.S.

Strishkov, V. V. 1973. "Soviet Coal Productivity: Clarifying the Facts and Figures." *Mining Engineering*, May, 1973.

Stugarev, A. S. 1976. *Prognozirovanie razvitiia ugol'noi promyshlennosti* [Forecasting Development of the Coal Industry]. Moscow.

Syrovarov, A. 1966. *Finansovaia rabota FZMK po gosudarstvennomu sotsial'-nomu strakhovaniiu* [Financial Work of Factory Committees for State Social Insurance]. Moscow.

Tartakovskii, G. S. 1978. *Ekonomika proizvodstva i effektivnost' ispol'zovaniia trub dlia magistral'nykh gazoprovodov* [Economics of Production and Effectiveness of Use of Pipe for Gas Transmission Lines]. Moscow.

Tolkachev, A. S., and I. M. Denisenko, eds. 1971. *Osnovnye napravleniia nauchno-tekhnicheskogo progressa* [Main Directions of Scientific Progress]. Moscow.

Treml, V. G. 1972. *The Structure of the Soviet Economy*. New York: Praeger.

Treml, V. G., and D. M. Gallik. 1973. *Soviet Studies on Ruble/Dollar Parity Ratios*. U.S. Department of Commerce, FDAD, Foreign Economic Reports. Washington.

TsNIEIUgol' (Central Economic Research Institute for Coal). 1976. *Transportirovanie uglia po truboprovodam* [Transporting Coal by Pipeline]. Moscow.

TsSU UkrSSR (Central Statistical Administration of the Ukrainian SSR). *Narodnoe khoziaistvo Ukrainskoi SSR.*

TsSU SSSR (USSR Central Statistical Administration). 1972. *Transport i sviaz'* [Transport and Communication]. Moscow.

TsSU SSSR. *Narodnoe khoziaistvo SSSR* [National Economy of the USSR].

Ulianov, I. A. 1972. *Ugli SSSR* [Coals of the USSR]. Moscow.

Umanskii, L. M., and M. M. Umanskii. 1974. *Ekonomika neftianoi promyshlennosti* [Economics of the Oil Industry]. Moscow.

Umanskii, L. M. 1962. *Puti snizheniia sebestoimosti v neftedobyvaiushchei promyshlennosti* [Ways of Reducing Costs in the Oil Extraction Industry]. Moscow.

U.S. Bureau of Mines. 1972. *Cost Analysis of Model Mines for Strip Mining of Coal in the United States*. Information Circular 8535. Washington.

U.S. Bureau of Mines. 1975. *Long Distance Coal Transport: Unit Trains or Slurry Pipelines*. Information Circular 8690/1975. Washington.

U.S. Bureau of Mines. 1976. *Coal Mine Equipment Forecast to 1985*. Information Circular 8710. Washington.

U.S. Bureau of Mines. *Minerals Yearbook.*

U.S. Congress: House. 1957. Committee on Interior and Insular Affairs. *Hearings*, 13–22 February 1957. Washington, D.C.

U.S. Congress: House. 1975. Science and Technology Committee. *U.S.-USSR*

Cooperative Agreements in Science and Technology, Hearings, 18–20 November 1975. Washington, D.C.

U.S. Department of Commerce. 1972. *U.S.-Soviet Commercial Relationships in a New Era*. Washington.

Ushakov, S. S. 1972. *Tekhniko-ekonomicheskie problemy transporta topliva* [Technical-Economic Problems of Transporting Fuel]. Moscow.

Vainshtein, B. S., and R. D. Takhenko. 1975. "Effektivnost' kompleksnykh vneshne-ekonomicheskikh investitsionno-proizvodstvennykh programm" [The Effectiveness of Complex Foreign Trade Investment Programs]. *Metody i praktika opredeleniia effektivnosti kapital'nykh vlozhenii i novoi tekhniki* [Methods and Practice for Determining the Effectiveness of Capital Investment and New Technology]. Vypusk 25, Moscow.

Vasil'ev, V. G. 1975. *Gazovye i gazokondensatnye mestorozhdeniia* [Gas and Gas-Condensate Fields]. Moscow.

Vilenskii, M. A. 1962. *Nauchno-tekhnicheskii progress i effektivnost' obshchestvennogo proizvodstva* [Scientific-Technical Progress and the Effectiveness of Social Production]. Moscow.

Vilenskii, M. A. 1963. *Elektrifikatsiia SSSR i razmeshchenie proizvoditel' nykh sil* [Electrification of the USSR and the Distribution of Productive Forces]. Moscow.

Wilczynski, Josef. "Atomic Energy for Peaceful Purposes in the Warsaw Pact Countries." Unpublished paper.

Zaleski, E. 1969. *Science Policy in the USSR*. Paris: OECD.

Zasiadko, A. F. 1959. *Osnovy tekhnicheskogo progressa ugol'noi promyshlennosti SSSR* [Foundations of Technical Progress of the Coal Industry of the USSR]. Moscow.

Zhimerin, D. G. 1978. *Energetika: nastoiashchee i budushchee* [Electric Power Today and Tomorrow]. Moscow.

Zhuravlev, V. P. 1971. *Analiz raboty ugol'nykh razrezov v 1966–1970gg* [Analysis of Operations of Coal Open-Pits in 1966–1970]. Moscow.

Zolotar'ev, T. L., and E. O. Shteingauz. 1960. *Energetika i elektrifikatsiia SSSR v semiletke* [Electric Power and Electrification of the USSR in the Seven Year Plan]. Moscow.

RUSSIAN LANGUAGE PERIODICALS

Atomnaia energiia [Nuclear Energy]

EKO, (Ekonomika i organizatsiia promyshlennogo proizvodstva—Economics and Organization of Industrial Production)

Ekonomicheskaia Gazeta [Economic Newspaper]

Ekonomicheskoe sotrudnichestvo stran-chlenov SEV [Economic Cooperation of Member Countries of the Council for Mutual Economic Cooperation]

Ekonomika gazovoi promyshlennosti [Economics of the Gas Industry]

Ekonomika neftianoi promyshlennosti [Economics of the Oil Industry]

Ekonomika i upravlenie ugol'noi promyshlennosti [Economics and Management of the Coal Industry]

Elektricheskie stantsii [Electric Power Stations]

Elektrichestvo [Electricity]

Energetik [Electric Power Worker]

Energetika i elektrifikatsiia [Electric Power and Electrification]

Energetika i transport [Electric Power and Transport]

Energomashinostroenie [Electric Power Machine Building]

Gazovaia promyshlennost' [The Gas Industry]
Geliotekhnika [Heliotechnology]
Izvestiia [News]
Izvestiia VNIIPT [News of the All-Union Scientific Research Institute for Direct Current]
Khimiia i tekhnologiia topliv i masel [Chemistry and Technology of Fuel and Oils]
Kommunist [The Communist]
Neftepererabotka i neftekhimiia [Petroleum Refining and Petrochemistry]
Neftianik [The Oil Worker]
Neftianoe khoziaistvo [The Oil Economy]
Planovoe khoziaistvo [Planned Economy]
Pravda [Truth]
Sobranie postanovlenii Pravitel'stva SSSR [Collection of Laws of the Government of the USSR]
Sotsialisticheskaia Industriia [Socialist Industry]
Stroitel'stvo truboprovodov [Pipeline Construction]
Teploenergetika [Thermal Power]
Truboprovodnyi transport: Itogi nauki i tekhniki [Pipeline Transport: Results of Science and Technology]
Ugol' [Coal]
Vestnik ANSSSR [Herald of the Academy of Sciences of the USSR]
Vestnik mashinostroeniia [Herald of Machine Building]
Voprosy ekonomiki [Questions of Economics]

ENGLISH LANGUAGE PERIODICALS
Annals of Nuclear Science and Engineering
Atomic Energy Review
Civil Engineering
Current Digest of the Soviet Press
Department of Energy Information
Electrical World
Electronics World
The Financial Times
Gas Turbine International
Gas Turbine World
Louisville Courier-Journal
Mining Engineering
Moscow Narodny Bank, *Press Bulletin*
New York Times
Nuclear News
Oil and Gas Journal
Science
Soviet News
Wall Street Journal

Index

A-1 nuclear reactor (Czech.), 154–55, 160

Academies of science, 34, 45–46, 49–50

Academy of Sciences of the Armenian SSR, 198

Academy of Sciences of the Kazakh SSR, 50

Academy of Sciences of the USSR: research and development, 19, 21, 27; oil and gas research, 45–46; electric power research, 48–50; geothermal energy research, 51, 191, 194

Academy of Sciences of the Uzbek SSR, 51–52, 198–99

Agriculture, 14

Air pollution. See Environmental problems

Aircraft engines, 214, 216, 218

Aleksandrov, A. P., 151, 155–56, 167

All-Union Geological Fund, 1–2

All-Union Oil and Gas Research Institute, 42

All-Union Research Institute of the Peat Industry, 52

All-Union Scientific Research Institute for the Construction of Pipelines, 226–27

All-Union Scientific Research Institute for Drilling Equipment, 42

Aluminum reduction industry, 5

Armenian nuclear power station, 141

Artsimovich, L. M., 158

Augers, 131

Austenitic steel, 74

Autoclave coal processing, 180

Automation, 65

Automobiles, 12

Baltic coast oil shale reserves, 6

Base-load electric power generation, 74–79

BelAZ-series trucks, 127, 128, 129, 135

Beloiarsk nuclear power station, 139, 140, 150, 151, 166

Belovo power station, 171, 172

Bernstein, L., 197

Bilibino nuclear power station, 140, 151

Black Mesa slurry pipeline (US), 172, 174

Block-design electric power generation, 74–79

BN-series nuclear power stations, 150–51, 164, 165, 166

Boiling-water nuclear reactors. See BWR-series nuclear reactors

Bratchenko, B. V., 107, 116

Breeder nuclear reactors: technological level, 142, 163–64; reprocessed fuel, 144–45, 159–63; plutonium fuel, 146; commercialization, 150–51; safety, 159, 168. See also BN-series nuclear power stations; Fast-breeder nuclear reactors

Briquets, 178

Bucket-wheel excavators, 104–5, 125–26, 136

Bulldozers, 104–5, 131

BWR-series nuclear reactors, 140–41

Cadiz-Eastlake slurry pipeline (US), 171, 173

Capital costs of energy, 25–26

Carryall scrapers, 104–5

Caspian oil fields, 4

Central Asia, gas resources, 3, 5

Central Boiler and Turbine Institute, 48

Channel-type light water graphite nuclear reactors. See RBMK-series nuclear reactors

Chemical machinebuilding, 44–45

Chemistry Institute (Estonian Acad. of Sci.), 52

Cheremkhovo coal field, 121

Cherepovets power station, 74, 75

Chernobyl nuclear power station, 141, 149

Chukhanov, Z. F., 177

Coal: resources, 1–2; production, 10; exports, .11; heat rate, 68; gas-steam combine turbines, 95; slurry transport, 170–75; gasification, 176–77; complex processing, 177–80

Coal conversion, 241–42

Coal mining: management, 20; use of resources, 20; development, 23, 24; labor and planning, 26; research and development, 30–31, 40–41, 43, 46–47, 54, 238–39, 246–47; machinery, 40–41, 46, 99–137, machine building, 46, 115–16; investment, 100, 102, 104–5; US-USSR compared (tables), 100–1, 104–5; loading machines, 101; technology imports, 203. See also Strip mining

Co-generation of energy, 15–16

Coke, 11, 178

Coking coal, 116–17

Combined cycle electric power generation, 64, 94–95

"Complex problems" of five-year plans, 28

Complex processing of coal, 177–80

Compressors for gas pipelines, 206, 207–9, 212–20, 227–29

Condensing power stations, 71, 74–79, 80

Condensing turbines, 63

Construction industry: energy use, 14

Container pipelines, 174

Conveyor transport, 117–18

"Coordination plans," 28

Cost-effectiveness, 32

Council for the Study of Productive Forces, 51

Council of Ministers, 19

Council on Reprocessing of Spent Fuel, 160

Current Sources, Research Institute of, 52

Czechoslovakia: A-1 nuclear reactor, 154–55; and USSR nuclear technology, 160–61; nuclear technology, 165

Demonstration projects, 35

Department of Chief Designer, 37

Design bureaus (KB), 37, 42–43, 46, 47–48

Design work, 36–37

DET tractor-dozer, 131

Diesel locomotives, 13, 126

Diffused-gas turbine, 89

Dmitrovgrad nuclear power station, 140

Dnepr coal mines, 117, 125

Dombrovskii, N. G., 124

Donbass coal basin, 111

Donbass-series wide-web mining combine, 109–10

Donets coal basin, 2

Donets Coal Institution, 46

Donetsk machine building plant, 125

Dongiprouglemash, 109–10

Dragline excavators, 122–25, 134, 136

Drilling equipment, 131–32

DU-1 narrow-web combine, 111

Dumpcars, 26, 130

Dvorov, M., 193

Eastern Coke-chemical Institute, 179

Educational Institutes, 34

"Effective" energy plan, 23–24

EGL-series mechanical shovels, 121

Eighth Five Year Plan (1966–1970), 28, 29, 117

EKG-series mechanical shovels, 120–21, 130

Ekibastuz-Center electric line, 182–83

Ekibastuz coal basin: production, 2, 23; bucket-wheel excavators, 125; electrical generation, 181–82

Electric locomotives, 101, 126, 130

Electric motor gas compressors, 209, 217

Electric power: exports, 11; research and development, 40–41, 47, 48–50, 59–60, 95–98, 242–43; technology export, 203

Electric power generation: peat fuel, 5–6; energy consumption, 13–16; natural gas, oil fuels, 16; thermal generation, 62–98; USSR-US compared (table), 66; equipment, 70–74; installed capacity (table), 72; base-load block condensing stations, 74–79; heat and power combines, 79–84; load variations, 86; magnetohydrodynamic generation, 184–91

Electric Power Institute (Latvian Acad. of Sci.), 49

Electric Power Institute (Minenergo) (ENIN), 34, 52, 198

Electric Power Research Institute (US), 157

Electric power stations, 6, 73, 75, 78, 89, 91, 94, 141, 171, 172, 178, 192

Electric power transmission: equipment, 47–48; research and development, 60; from Siberia, 181–84

Electric pumps for oil industry, 221–25

Electron-beam fusion, 157

Elektroset'proekt, 21

Employment: in research and development, 38–57; in coal mining, 100, 104–5

Energy: resources, 1–12; consumption, 8–12; production (table), 10; trade, 10, 11, 24–25; conservation of, 17, 53, 55; management, 18–25; planning, 21; novel sources, 202

Energy Research and Development Ad-

ministration (US), 29–30
Environmental problems: oil shale recovery, 6; damage to land, 7; research and development, 53, 55, 58; electricity generation, 65; nuclear power, 158–59; magnetohydrodynamic generators, 187
ERG-series bucket excavators, 125–26
Ermakov power plant, 75
ESh-series dragline excavators, 122–24
ETKh-series coal converters, 178, 179
EVG-series mechanical shovels, 120, 122
Excavators, 119–26, 133
Experimental-design work (OKR), 36, 43

Fast breeder (FBR-series) nuclear reactors, 140–41
Finland, nuclear technology, 160, 162
Firewood, 6, 9, 10
Five year plans, 27–28. See also Seventh, Eighth, Ninth, and Tenth Five Year Plans
Fluidized bed coal processing, 179
Forecasting: energy demands, 22–25; nuclear research, 145–48
Foreign exchange, 24, 230
Fossil fuels: research and development, 53, 54, 59; electric power generation, 62–98. See also Coal; Lignite; Natural gas; Oil; Oil shale; Peat
Freon-12, 193–94
Front-end loaders, 104–5
Fuel: production by type (table), 10; consumption by sector (table), 15; rationing, 17; electric power generation, 63. See also specific fuels
Fuel oil, 95
Fusion nuclear power, 155–58

Gas. See Natural gas
Gas condensate, 2–3
Gas-cooled nuclear reactors, 153–54
Gas Industry, Research Institute of the, 43
Gas Institute (Ukrainian Acad. of Sci.), 45
Gas turbine compressors, 209, 212–14, 215, 218, 228
Gas turbines: electric power generation, 63, 88–92; US, 91–92; steam turbine combination, 94–95
Gasification, 175, 176–77
Geliotekhnika (solar energy journal), 51–52

General Electric (US) gas compressors, 218–19, 228
Geological Sciences Institute (Armenian Acad. of Sci.), 51
Geology, research and development, 40–41, 53
Geology and Production of Mineral Fuels Institute, 42–43
Geophysics Institute (Georgian Acad. of Sci.), 50, 51
Geothermal energy: resources, 7; research and distribution, 51, 53, 54, 191–96; power stations, 191–96
German Democratic Republic: bucket-wheel excavators, 125–26; nuclear power, 155
German Federal Republic, nuclear technology, 164
Gidroproekt. See Project-making organizations
Giprotsentroshakt, 135
Giprouglemash, 113, 114
GK-series gas turbine compressors, 214
Golovnoi obrazets (prototype equipment), 96–97
Golovnye institutes, 47
Gosplan, 19, 21, 27, 44–45
GPA-series gas compressors, 216, 218
Great Britain, nuclear equipment, 165
Grigoriants, A., 152, 159
Gross National Product, 9
GT-series gas turbine compressors, 213
GTN-series gas compressors, 216–17, 228
Gubkin Institute, 45, 46

Heat: by-product of power generation, 16; supply by nuclear reactors, 144, 151, 152
Heat and power combines, 79–84, 186
Heat content of coal, 100
Heat load of heat and power combines, 81–82
Heat rate: electric power generation, 62, 66–68; steam condensing generators, 77–78; heat and power combines, 80–81; gas turbines, 91–92
Heat turbines, 63
Heliotechnology, 198–202
High Temperature Institute (Moscow Power Inst.), 49, 185
High-voltage power transmission, 181–84
Household consumption of energy, 14
HTGR-series nuclear reactors, 153
Hybrid nuclear reactors, 156–57

Hydraulic coal transport. *See* slurry pipelines

Hydraulic strip mining, 117

Hydroelectric power: resources, 4–5; share of energy sources, 12; capital costs, 26; peak demand generation, 88

Hydrogenation of coal, 175, 179–80

Hydrogeology and Geothermics, Commissions on, 191

Hydropower, 10

IaKZ-series dump trucks, 127

Iaroslav heavy truck plant, 127

Ignalina (Lithuanian) nuclear power station, 141, 145

"Import equivalent," 226

Import of energy, 10, 24–25

Industry, energy used, 13–15, 16, 17

Inskoe coal mine, 171

Installed capacity of power stations, 69–70

"Integrated systems approach," 32–33

Ioffe Physico-technical Institute, 50

Irrigation projects, 5

Irsha-Borodinsk coal mine, 125

Iterative hierachical modeling, 145–46

Iubileinaia coal mine, 171

Izhorsk machine building plant, 120, 121

Japan: uranium purchases, 160; control equipment supplier, 164–65

K-52M narrow-web combine, 111, 113

Kalinin nuclear power station, 141

Kansk-Achinsk coal: gasification of lignite 176–77; liquid fuel processing, 177–80

Kansk-Achinsk coal basin: production, 2, 23, 24; strip mines, 117; mechanical shovels, 122; dragline excavators, 124; hydraulic coal transport, 171; slurry pipeline projects, 173–75; lignite transport, 175–76; electric power generation, 181–82

Karaganda coal basin, 2

Kazakh SSR, natural gas resources, 3

Kerogen, 6, 52

Khar'kov Turbine Factory, 74, 75, 164

Khar'kov Turbogenerator Plant, 89

Kiev-1 magnetohydrodynamic generator, 185

Kirillin, V. A., 180, 185, 189, 199–200

Kislaia Bay tidal power station, 197

KM-87 narrow-web mining complex, 112–14

Kola nuclear power station, 139, 140

Komi ASSR, natural gas resources, 3

Konstruktorskaia rabota (design work), 36–37

Krasnodar power plant, 91

Krasnoiarsk heat and power combine, 178

KrAz-series dump trucks, 127

Krivoi Rog regional power station, 73

Krylov, A. P., 221

Krzhizhanovskii Power Institute, 177, 178, 179, 185

Kupianskaia gas compressor station, 216

Kurchatov Institute, 50, 155, 157, 185

Kursk nuclear power station, 141

Kuzbass coal basin, 121, 171–74

Kuznetsk coal basin, 2

Labor: energy planning component, 26; in research and development, 57; power plants, 64–65; coal mining, 98, 106–8; blast-hole drilling, 131–32

Laboratory of Geothermics and Hydrogeochemistry (USSR Acad. of Sci.), 51

Lalaiants, A., 174

Lebedev Physics Institute, 50

Leningrad Electrotechnical Institute, 49–50

Leningrad Metal Factory, 74, 75, 89, 94

Leningrad Mining Institute, 47, 51

Leningrad nuclear power station, 140, 142, 145, 164, 167

Leningrad Polytechnical Institute, 49

Light-water graphite nuclear reactors, 140-41

Lignite, 175. *See also* Kansk-Achinsk coal

Liquid hydrocarbons, 6

Liquifaction, 175

Lithuanian nuclear power station, 141, 145

Locomotives, 12–13, 101, 104–5, 126, 130

Long-wall coal mining, 106, 108–9

Loviisa nuclear power plant (Finland), 162

Low-grade fuels, 7, 11, 12

LWGR-series nuclear reactors, 140–41

Magnetic confinement nuclear reactor, 157

Magnetohydrodynamic power generation, 184–91

Makhachkala geothermal power station, 192

Malakovskii machine building factory, 111, 113

Marine gas turbines, 216
MAZ-series trucks, 127
Mechanical shovels, 119–22
Meyerhoff, A. A., 223
Mezen tidal power location, 196, 197
Mine-mouth electric power stations, 181
Mineral Fuels, Institute for Research on, 177
Ministries, research and development, 30–31, 40–41
Ministry for the Construction of Oil and Gas Enterprises (Minneftegazstroi), 40–41
Ministry of Agriculture, 195
Ministry of Chemical and Petroleum Machinebuilding (Minkhimmash), 40–41, 44–45
Ministry of Control Equipment (Minpribor), 18–19
Ministry of Electric Equipment (Minelektrotekhprom), 18, 40–41, 47–48
Ministry of Electric Power and Electrification (Minenergo): organization, 18; coal research and development, 30–31; project-making institutions, 35–36; research and development expenditures, 40–41, 47; geothermal research, 51; electric power generation, 82, 83; peaking problems, 92; pumped water storage generation, 93; electric power research and development, 98; nuclear research, 166–67; coal combine design, 178–79; magnetohydrodynamic power generation, 189; geothermal research, 194–95; tidal power research, 235; technological advances, 237, 247
Ministry of Ferrous Metals (Minchermet), 19
Ministry of Geology (Mingeo): organization, 18, 20; research and development, 39–42; geothermal research, 51, 191, 193
Ministry of Heavy, Energy, and Transport Machine Building (Mintiazhmash), 19, 20, 44, 115–16
Ministry of Oil Extraction (Minneft'), 18, 20, 40–41, 42–43
Ministry of Oil Refining and Petrochemicals (Minneftekhim), 18, 40–41, 43–44
Ministry of Petroleum Machine Building (Minneftemash), 19
Ministry of Power Equipment (Minenergomash), 40–41, 48
Ministry of the Coal Industry (Minugol'): organization, 18, 20; research

and development, 30–31, 40–41, 46–47, 116
Ministry of the Gas Industry (Mingaz): organization, 18, 20; research and development, 40–41, 43; geothermal research, 51, 194, 195; gas pipelines (tables), 205, 206
Ministry of Water Economy, 201
Minsk automobile factory, 127
MK-1 narrow-web mining complex, 114
Modeling: energy policies, 21–25; research and development, 30–33; nuclear research, 145–48
Moscow coal basin, 111, 117
Moscow Electric Power Institute, 49
Motor fuel, 177

Narrow-web mining combines, 109, 111–16
National Science Foundation (US), 56–57
National security, and plutonium, 159–63
Natural gas: resources, 2–3, 176; production, 10; electric power generation, 16; role in energy planning, 23, 24; exploration for, 39–42; magnetohydrodynamic generators, 185–86
Natural gas industry: management, 20; research and development, 36, 40–41, 43, 45–46, 53, 54; technology imports, 203, 209–21
Natural gas pipelines, 204–21. See also Compressors for gas pipelines
Nazarovo coal mine, 124
Nazarovo power station, 78
Nebit-Dag power station, 89
Neporozhnyi, P. S., 47, 83, 150, 162, 193, 197
Neva Machine Plant, 89
Nevinnomysskaia power station, 94
Nevskii machine building plant, 213
Ninth Five Year Plan (1971–1975): oil and gas, 45–46; gas turbines, 90; strip mining, 117; dragline excavators, 124; heavy trucks, 128; nuclear reactors, 142; electrical transmission lines, 183; gas compressors, 217
NK-12-ST aircraft engine, 214
Novo - Kramatorsk machine - building plant, 122, 126
Novo-Voronezh nuclear power station, 139, 140, 162, 166
Nuclear fuel: resources, 6–7; reserves, 143–44; supply, 144, 146; product of breeder reactors, 149, 150, 156; reprocessing, 159–63

Nuclear Physics Institute (Belorussian Acad. of Sci.), 50, 152

Nuclear power: energy alternative, 7; production, 10; development, 11–12; dependence on, 23, 24; problems, 25; research and development, 40–41, 50–51, 53, 54, 58–59, 143–59, 165–69; 241; electric power generation, 62–63; growth, 139; aid in geothermal energy recovery, 193–94

Nuclear power stations: research and development, 35–36; heat and power combine design, 79, 83; peak electrical demand, 88; use with pumped water stations, 93; individual stations, 138, 139, 140, 141, 142, 145, 149, 150, 151, 162, 164, 166, 167; characteristics (table), 140–41; BN-series, 150–51, 164, 165, 166

Nuclear reactors: installed in power stations (table), 140–41; export, 203. See also Breeder; BWR-series; HTGR-series; LWGR-series; RBMK-series; and VVER-series nuclear reactors

Nuclear Reactors Research Institute, 50

Nuclear waste, 161–62

Obninsk nuclear power station, 138, 139, 166

Offshore natural gas, 3, 4

Oil: reserves, 3–4; production, 9, 10, 21, 36; exports, 11; generation of electricity, 16; use by transport, 16; production planning, 21; role in energy plans, 23, 24; exploration, 39–42; research and development, 40–41, 42–44, 45–46, 53, 54

Oil industry: machine building, 44–45; technology imports, 203; submersible electric pumps, 221–25

Oil pipelines, 210

Oil shale, 6, 10, 52, 175

OMKT narrow-web mining complex, 111, 112

Open-pit mining. See Strip mining

"Optimal" energy plan, 23

Orenburg gas field, 3

Orudzhev, S. A., 216

Paratunka geothermal field, 193

Paratunka geothermal power station, 193

Pauzhetsk power station, 191, 192

Peak electrical loads, 85–89

Peat: resources, 5–6; production, 10; research and development, 52; electrical

power generation, 62

Peat Institute (Belorussian Acad. of Sci.), 52

Perlitic steel, 74

Petrochemicals, 43–44

Petroleum. See Oil

Petropavlovsk-Kamchatskii geothermal power station, 193

Petrosiants, A. M., 50, 143, 144, 150, 161

Physico-technical Institute (Turkmen Acad. of Sci.), 50, 52, 199

Physico-technical Institute (Ukrainian Acad. of Sci.), 50

Pipe: natural gas, 204–21; production, 211; research and development versus importation, 225–27

Piston gas compressors, 209, 217

Plutonium: supply, 144, 146; breeder reactor source, 149, 156; reprocessing, 159–63

Pollution. See Environmental problems

Power and Hydraulics Institute (Armenian Acad. of Sci.), 198

Power Institute (ENIN). See Electric Power Institute

Pressurized-light-water nuclear reactors. See VVER-series nuclear reactors

Price policy, as energy conservation method, 17

Pridniepr power station, 75, 172

Problems of Deep Fields Institute, 45

"Problems" of five year plans, 28

"Programs" of five year plans, 28, 29

Project-design organizations (PKO), 21, 36, 44

Project-making organizations: functions, 35–36, 167; Minenergo, 47; pumped water storage, 92; nuclear power station design, 166–67; tidal power development, 197

Promenergoproekt, 179

Prototype equipment, 96–97

Pumped water storage power generation, 63, 88, 92–93, 196

Pumps: for slurry pipelines, 173; submersible electric, 221–25

Railroads, 13, 118–19, 126, 130, 170. See also Locomotives

"Rational" energy plan, 24

RBMK-series nuclear reactors: installations, 140–41; Leningrad power station, 142, 145; development, 148–50, 163; Czech equipment, 165; design, 166–67; problems, 243

Refrigeration Machinery Institute, 193

Relevance tree for nuclear power forecasting, 146, 147
Reprocessing centers for nuclear fuel, 161
Research and development: ministries' control over, 19–21; planning, 27–33; expenditures (table), 40–41; US, 53–60; USSR-US compared, 57–60; import alternative, 225–30; assessments of USSR programs, 231–49. *See also specific fields of research*
Research and project-making institutions (NIPI), 36, 42–43
Reserve electrical capacity, 63, 69
Rolls Royce Avon engines, 219
Room and pillar coal mining, 103–4
Rudakov, L. I., 157
Rudins, G., 188, 190

Safety: research and development, 53, 155; neglect of, 58; strip mines, 116; nuclear power, 142–43, 158–59, 167–68
Scientific Council on Geothermal Problems, 51, 191
Scientific-production associations (NPO), 43
Scientific Research Institute for Shale, 52
Scientific-research institutes (NII), 35, 37, 42–43, 44, 46, 47–48
Scientific-research work (NIR), 36, 49
Scientific workers, 38–57
Scrapers, 131
SE-3 mechanical shovel, 120
Secondary-fluid geothermal power station, 192–93
Semicoke, 178
Semipeak electric power equipment, 93–94
Seventh Five Year Plan (1959–1965): gas turbines, 89; heavy trucks, 128
Shale. *See* Oil shale
Shashin, V. D., 222
Shatskaia power station, 89
Shatura electric power station, 6
Shchekino gasification plant, 176
Sheindlin, A. E., 185, 188
Shevchenko nuclear power station, 140
Shuttle cars, 101
Siberia: hydroelectric resources, 5; natural gas reserves, 3, 4
Siberian Energy Institute, 21, 49
Siberian nuclear power station, 78, 139, 140, 161
Siberian transmission line, 183–84

Single bucket excavators, 103, 104–5
Slurry pipelines, 170–75
Smolensk nuclear power station, 141, 149
Sodium coolant in nuclear reactors, 152
Solar cells, 199–200
Solar Energy Institute, 201
Solar furnace, 199
Solar power, 7, 51–52, 53, 54, 198–202
Solid fuel conversion, 175–80
Solvent refining, 175
South Ukrainian nuclear power station, 141
Space heating, 144, 151
Starodubtsev Physico-technical Institute, 51–52, 198–99
State Committee for the Peaceful Uses of Atomic Energy, 19, 34, 50, 166–67
State Committee for Science and Technology, 27, 28, 30, 31, 228
State Secrets Act, 3
Steam condensing power generation, 71, 74–79, 80
Steam locomotives, 12–13, 126, 130
Steam pressure, and electric power generation, 64
Steam shovels. *See* mechanical shovels
Steam turbines, 94–95
Steam turbogenerators, 70
Steel: in turbines, 74; boilers, 76; strip mine equipment, 134; pipe, 212
Strip mines: environmental problems, 7; USSR-US compared, 100, 102–3, 104–5; machinery, 116–32; research and development, 135–36
Styrikovich, M. A., 49, 91
Submersible electric pumps, 221–25
Sulfurous oil, 186
Superconducting magnet, 187
Sverdlovsk power station, 178
Synthetic liquid fuels, 177–80
"Systems approach" to forecasting, 32–33

T-20 magnetic confinement reactor, 157
Tashkent Communications Institute, 52
"Tasks" of five year plans, 28
Technology transfer, 203–30
Tenth Five Year Plan (1976–1980): energy goals, 21; research and development, 28; steam-condensing power stations, 78–79; semipeaking electrical equipment, 93–94; pumped water storage research, 93; gas-steam combines, 95; strip mining, 117; heavy trucks, 128; nuclear reactors, 142, 151–

52; nuclear fuels, 143–44; electrical transmission lines, 182–83; solar energy, 200–1

Teploenergoproekt (Minenergo): organization, 35; nuclear research, 166; coal conversion, 178; geothermal power, 192

Teplofikatsiia (heat and power combines), 79–84, 186

Teploproekt, 47

Tiumen' oblast', natural gas reserves, 3

Thermal-contact coal processing, 178–79

Thermal-neutron gas-cooled nuclear reactors, 153

Thermal power generation, 62–98

Thermophysics Institute (USSR Acad. of Sci.), 49

Tidal energy, 7, 196–98, 235

TKKU-series coal processors, 179

Tokamak-series nuclear reactors, 156

Toretskii machine building plant, 113

Tractors, 101

Transport system of strip mining, 117–19

Transportation: costs and energy planning, 7, 25; energy use, 12–13, 15, 16

Transportless strip mining, 117, 118

Troitsk (Siberian) nuclear power station, 78, 139, 140

Trolleivoz diesel-electric truck, 129

Trucks, 118–19, 126–30

Turbodrill, 236

Turboprop aircraft engine, 214

U-25 magnetohydrodynamic generator, 35, 185, 186, 187, 188, 189, 242–43

Uglegorsk power station, 78

Ukrainian SSR, research and design enterprises, 37

Underground coal mining, 100, 102, 103–7, 108–16

United States: energy research and development, 29, 53–60, 244, 245; environmental problems, 55; electric power generation, 59–60, 66; gas turbines, 91–92; underground coal mining, 100; strip mining, 100, 102–3, 104–5; slurry pipelines, 171; magnetohydrodynamic generator research, 187–88; solar energy, 200, 201; gas pipelines, 206–8; gas compressors, 218–19; submersible electric pumps, 222–24

Ural Heavy Machine Building Plant (Uralmashzavod), 120, 123–24

Uranium: resources, 6–7; reserves, 143–44; U-233, 150; sale to Japan, 160

Uranium nuclear reactors, 154–55

Velikov, E., 156

Vilenskii, M. A., 83

Volga basin: hydroelectric power, 5; oil shale reserves, 6

Volga-Ural oil region, 9, 222

Voltage regulation, 88

VVER-series nuclear reactors: installations, 139–41; research and development, 145, 148–50, 163; in Czechoslovakia, 155, 160; plutonium fuel, 160; containment structure, 162; Czech equipment, 165; problems, 243

Wage bill, 40–41, 42–57

Walking dragline excavator, 122–25, 136

Wall cutters, 101

West Ukrainian nuclear power station, 141

Wheeled loaders, 131

Wide-web mining combines, 109–11

Wood. See Firewood

Zagorsk pumped water storage plant, 93, 198

Zaporozh'e power station, 78

DATE DUE

DEMCO 38-297